JAMESTOWN PUBLISHERS

THE WILD SIDE

Weird Science

JAMESTOWN PUBLISHERS

a division of NTC/Contemporary Publishing Group
Lincolnwood, Illinois USA

Photo Credits
Biological Photo Service: 22, 54; FPG: 68/Paul Popper, 74/G. Randall, 86/Terry Qing; 88/Telegraph Colour Library; The Image Bank: 12/Patricia A. McConville, 60/JPH Images; NASA: 2, 16; Stock Montage: 34/Charles Walker, 66

Cover illustration: Tim Jessell

ISBN: 0-89061-798-8

Published by Jamestown Publishers,
a division of NTC/Contemporary Publishing Group, Inc.,
4255 West Touhy Avenue,
Lincolnwood (Chicago), Illinois 60646-1975 U.S.A.

Manufactured in the United States of America.

8 9 0 DBH 9 8 7 6 5 4 3

Contents

GROUP THREE

To the Teacher

INTRODUCTION

Biology, geology, psychology, physics, chemistry—to many students, naming these branches of science calls forth little more than a yawn. Stories in *Weird Science*, however, go far beyond naming areas of study. They focus on details of natural events, insect behavior, human experiences, and other subjects of scientific investigation that fascinate even as they shock, repel, or mystify readers. After becoming aware of such puzzling happenings and fantastic theories, no reader can hold on to the attitude that science is dull or for the specialist.

Weird Science provides subject matter for thoughtful interpretation and discussion, while challenging your students in four critical reading categories: main idea, important details, inferences, and vocabulary in context. *Weird Science* can also help your students to improve their reading rates. Timed reading of the selections is optional, but many teachers find this an effective motivating device.

Weird Science consists of fifteen units divided into three groups of five units each. All the stories in a group are on the same reading level. Group One is at the fourth-grade reading level, Group Two at the fifth, and Group Three at the sixth, as assessed by the Fry Formula for Estimating Readability.

HOW TO USE THIS BOOK

Introducing the Book. This text, used creatively, can be an effective tool for teaching certain critical reading skills. We suggest that you begin by introducing the students to the contents and format of the book. Discuss the title. What is "weird" about science? Ask the students to recall topics they have read about or discussed in science classes that have stirred their curiosity or imagination. Read through the table of contents as a class to gain an overview of the topics that will be covered.

The Sample Unit. Turn to the Sample Unit on pages 1–7. After you have examined these pages yourself, work through the Sample Unit with your students, so that they have a clear understanding of the purpose of the book and of how they are to use it.

The Sample Unit is set up exactly as the fifteen regular units are. The introductory page includes a photograph or illustration accompanied by a brief introduction to the story. The next two pages contain the story and are followed by four types of comprehension exercises: Finding the Main Idea, Recalling Facts, Making Inferences, and Using Words Precisely.

Begin by having someone in the class read aloud the introduction. Then give the students a few moments to study the picture. Ask them to predict what the story will be about. Continue the discussion for a minute or so. Then have the students read the story. (You may wish to time the students' reading in order to help them increase their reading speed and improve their comprehension. Students can use the Words-per-Minute tables located on pages 110-112 to help them figure their reading speed.)

Work through the sample exercises as a class. At the beginning of each exercise are an explanation of the comprehension skill and directions for answering the questions. Make sure all the students understand how to complete the four different types of exercises and how to determine their scores. The correct answers for the exercises and sample scores are printed in lighter type. Also, explanations of the correct answers are given for the sample Finding the Main Idea and Making Inferences exercises to help the students understand how to think through these question types.

As the students work through the Sample Unit, have them turn to the Words-per-Minute tables (if you have timed their reading) and the Reading Speed and Critical Reading Scores graphs on pages 113 and 114 at the appropriate points. Explain the purpose of each feature and read the directions with the students. Be sure they understand how to use the tables and graphs. You may need to help them find and mark their scores for the first several units.

After they finish the Sample Unit, read and discuss the To the Student introduction on page 8 with the class.

Timing the Story. If you choose to time your students' reading, explain your reason for doing so: to help them track and increase their reading speed.

One way to time the reading is to have all the students in the class begin reading the story at the same time. After one minute has passed, write on the chalkboard the time that has elapsed. Update the time at ten-second intervals (1:00, 1:10, 1:20, etc.). Tell the students to copy down the last time shown on the chalkboard when they finish reading. They should then record this reading time in the space designated after the story.

Have the students use the Words-per-Minute tables to check their reading rates. They should then enter their reading speed on the Reading Speed graph on page 113. Graphing their reading times allows the students to keep track of increases in their reading speed.

Working Through Each Unit. When the students have carefully completed all parts of the Sample Unit, they should be ready to tackle the regular units. Begin each unit by having someone in the class read aloud the introduction to the story, just as you did in the Sample Unit. Discuss the topic of the story and allow the students time to study the illustration.

Then have the students read the story. If you are timing the reading, have the students enter their reading time, find their reading speed, and record their speed on the graph after they have finished reading the story.

Next, direct the students to complete the four comprehension exercises without looking back at the story. When they have finished, go over the questions and answers with them. Have the students grade their own answers and make the necessary corrections. Then have them enter their Critical Reading Scores on the graph on page 114.

The Graphs. Students enjoy graphing their work. Graphs show, in a concrete and easily understandable way, how a student is progressing. Seeing a line of progressively rising scores gives students the incentive to continue to strive for improvement.

Check the graphs regularly. This will allow you to establish a routine for reviewing each student's progress. Discuss with each student what the graphs show and what kind of progress you expect. Establish guidelines and warning signals so that students will know when to approach you for counseling and advice.

RELATED TEXTS

If you find that your students enjoy and benefit from the stories and skills exercises in *Weird Science,* you may be interested in *Crime and Punishment, Extreme Sports, Angry Animals, Bizarre Endings,* and *Total Panic:* five related Jamestown texts. All feature high-interest stories and work in four critical reading comprehension skills. As in *Weird Science*, the units in those books are divided into three groups, at reading levels four, five, and six.

Sample Unit

Do you want to walk on the moon someday? One of the few humans who has enjoyed this adventure is astronaut Buzz Aldrin. He is shown here next to the American flag that the Apollo 11 landing crew set up on the moon in July, 1969. But if you're set on visiting the moon, perhaps you'd better hurry. If Professor Alexander Abian has his way, the moon won't be around very long. Abian wants to blow up the moon! Why? He thinks it will improve weather on the earth. Not many people—not even many scientists—agree with him.

SHOULD WE BLOW UP THE MOON?

Most people like the moon just the way it is. They write poems about it. They sing love songs to it. They hold hands under it. But Alexander Abian has a scheme that would change all that. He wants to blow up the moon!

Abian is a mathematics professor at Iowa State University. He has a bold plan. First, he wants to send some astronauts to the moon. They would drill a huge hole in the moon's surface. Into this hole they would tuck some nuclear bombs. After the astronauts are safely out of the way, someone back on Earth would push a remote control button. One second later, the moon would be blown to bits.

Why does Abian want to do this? He thinks it would improve the earth's weather. With the moon out of the way, he says, there would be no more blizzards in the Rocky Mountains. There would be no killer typhoons in Asia. Summer heat waves in New York City would end. So, too, would droughts in Africa. Not only would bad things end, but good things would start. According to Abian, the deserts and arctic regions would bloom. After we blow up the moon, says the professor, we would have pleasant weather all year long.

What does the moon have to do with snowstorms in Denver or floods in Bangladesh? Plenty, says Abian. The moon's gravity pulls on the earth. That tug keeps the earth tilted at a 23½-degree angle. And that's the problem. It is this tilt that gives us our seasons. The side of the earth tilted toward the sun has summer and sweltering weather. The side tilted away from the sun has winter and chilling cold.

Now suppose we blow up the moon. According to Abian, the earth would then lose its 23½-degree tilt. The amount of sunlight would no longer change with the seasons. It would be the same all year long. "Perpetual spring!" promises Abian.

So why haven't we blown up the moon? Most people like having it around. More than a dozen countries like it so much they have put it on their national flags. Abian understands that. So he has come up with a second plan. He says we could try having *two* moons. We could "bring a moon from Mars." It could be put on the other side of Earth from the first moon. That way, its pull would balance off the pull of the original moon. Now the Earth would have two moons but no tilt!

There is another serious problem with blowing up the moon. True, it might get rid of the earth's tilt. But such a change might cause massive earthquakes. David Taylor of Northwestern University observes, "[Abian] would destroy civilization. But we'd have great weather." Thomas Stix of Princeton adds that most scientists wouldn't touch Abian's idea "with a ten-foot pole."

Such talk doesn't bother Abian. He wants to shake things up. Why, he asks, do we have to accept the solar system the way it is? Why can't we move things around? Abian has some other ideas, as well. He would like to change the orbit of Venus. It's too close to the sun, he says. Temperatures on Venus are a toasty 900°F. Abian thinks we should move Venus away from the sun. That would cool the planet and perhaps make it fit for human life. How does Abian recommend we move Venus? "We can shoot it with rockets," he suggests.

No one is holding his or her breath waiting for these things to happen. Even Abian knows that other scientists think his ideas are a bit strange. "I don't think [anything will happen] in my lifetime or in my children's lifetime," he says. "But I want to plant the seed."

If you have been timed while reading this selection, enter your reading time below. Then turn to the Words-per-Minute table on page 110 and look up your reading speed (words per minute). When you are working through the regular units, you will then enter your reading speed on the graph on page 113.

READING TIME: Sample Unit

_____________ : _____________

Minutes *Seconds*

How Well Did You Read?

- *The four types of exercises that follow appear in each unit of this book. The directions for each type of exercise tell you how to mark your answers. In this Sample Unit, the answers are marked for you. Also, for the Finding the Main Idea and Making Inferences exercises, explanations of the answers are given to help you understand how to think through these question types. Read through these exercises carefully.*

- *When you have finished all four exercises in a unit, use the answer key that starts on page 105 to check your work. For each right answer, put a check mark (✓) on the line beside the box. For each wrong answer, write the correct answer on the line.*

- *Find your scores by following the directions after each exercise. In this unit, sample scores are entered as examples.*

A FINDING THE MAIN IDEA

A good main idea statement answers two questions: it tells *who* or *what* is the subject of the story, and it answers the understood question *does what*? or *is what*? Look at the three statements below. One expresses the main idea of the story you just read. Another statement is *too broad*; it is vague and doesn't tell much about the topic of the story. The third statement is *too narrow*; it tells about only one part of the story.

Match the statements with the three answer choices below by writing the letter of each answer in the box in front of the statement it goes with.

M—Main Idea B—Too Broad N—Too Narrow

✓ [N] 1. Professors at Northwestern University and at Princeton disagree with the ideas of Alexander Abian.
[This statement is true, but it is too narrow. It doesn't suggest what Abian's ideas are about.]

✓ [M] 2. Mathematics professor Alexander Abian has proposed blowing up the moon as a way of improving weather on Earth.
[This is the main idea. It tells whom the story is about and what he did.]

✓ [B] 3. It's hard to tell whether some theories about the universe should be taken seriously.
[This statement is too broad. It doesn't tell which theory the story is about.]

15 Score 15 points for a correct *M* answer.
10 Score 5 points for each correct *B* or *N* answer.

25 TOTAL SCORE: Finding the Main Idea

B RECALLING FACTS

How well do you remember the facts in the story you just read? Put an *x* in the box in front of the correct answer to each of the multiple-choice questions below.

1. Abian's theory is that destroying the moon will
   - ____ ☐ a. cause Earth to tilt at a 23½-degree angle.
   - ____ ☐ b. cause massive earthquakes.
   - _✓_ ☒ c. improve weather on Earth.

2. The side of Earth tilted toward the sun has
   - _✓_ ☒ a. summer.
   - ____ ☐ b. winter.
   - ____ ☐ c. perpetual spring.

3. Abian also suggests moving
   - ____ ☐ a. the moon to a new orbit.
   - ____ ☐ b. Earth closer to Venus.
   - _✓_ ☒ c. a moon from Mars to orbit Earth.

4. One way to move Venus, Abian says, is to
   - _✓_ ☒ a. shoot it with rockets.
   - ____ ☐ b. set off nuclear bombs on it.
   - ____ ☐ c. use a ten-foot pole as a lever.

5. Abian expects his ideas to get serious attention
   - ____ ☐ a. next year.
   - ____ ☐ b. during his lifetime.
   - _✓_ ☒ c. in the distant future.

Score 5 points for each correct answer.

_25_ TOTAL SCORE: Recalling Facts

C MAKING INFERENCES

When you use information from the text and your own experience to draw a conclusion that is not directly stated in the text, you are making an *inference*.

Below are five statements that may or may *not* be inferences based on the facts of the story. Write the letter *C* in the box in front of each statement that is a correct inference. Write the letter *F* in front of each faulty inference.

C—Correct Inference F—Faulty Inference

_✓_ [F] 1. Most mathematics professors are more creative than scientists.
[This is a *faulty* inference. It makes a value judgment without supporting evidence.]

_✓_ [F] 2. Other professors criticize Professor Abian's ideas because they are jealous of him.
[This is a *faulty* inference. Other reasons are given.]

_✓_ [C] 3. As we humans gain power over nature, we must guard against unwise use of this power.
[This is a correct inference. Abian suggests using present technology to make vast changes.]

_✓_ [C] 4. Humans cannot live on or even explore Venus.
[This is a *correct* inference. The temperature of Venus is 900°F.]

_✓_ [C] 5. Professor Abian's main goal in proposing "corrections" to the universe is to get people to take a fresh look at things they take for granted.
[This is a *correct* inference. Abian says, "I want to plant the seed."]

Score 5 points for each correct *C* or *F* answer.

_25_ TOTAL SCORE: Making Inferences

D USING WORDS PRECISELY

Each numbered sentence below contains an underlined word or phrase from the story you have just read. Under the sentence are three definitions. One is a *synonym*, a word that means the same or almost the same thing as the underlined word: *big* and *large* are synonyms. One is an *antonym*, a word that has the opposite or nearly opposite meaning: *love* and *hate* are antonyms. One is an unrelated word; it has a completely *different* meaning than the underlined word. Match the definitions with the three answer choices by writing the letter that stands for each answer in the box in front of the definition it goes with.

S—Synonym A—Antonym D—Different

1. After the astronauts are safely out of the way, someone back on Earth would push a remote control button.

✓ [S] a. distant
✓ [A] b. close
✓ [D] c. powerful

2. Summer heat waves in New York City would end. So, too, would droughts in Africa.

✓ [A] a. floods
✓ [D] b. sicknesses
✓ [S] c. unusually dry spells

3. The side of Earth tilted toward the sun has summer and sweltering weather.

✓ [S] a. extremely hot and humid
✓ [A] b. very cold
✓ [D] c. cloudy

4. "Perpetual spring!" promises Abian.

✓ [D] a. dangerous
✓ [A] b. temporary
✓ [S] c. everlasting

5. How does Abian recommend we move Venus?

✓ [S] a. advise that
✓ [D] b. recall that
✓ [A] c. discredit the idea that

15 Score 3 points for a correct *S* answer.
10 Score 1 point for each correct *A* or *D* answer.

25 TOTAL SCORE: Using Words Precisely

- *Enter the total score for each exercise in the spaces below. Add the scores to find your Critical Reading Score. Then record your Critical Reading Score on the graph on page 114.*

Score	Exercise
25	Finding the Main Idea
25	Recalling Facts
25	Making Inferences
25	Using Words Precisely
100	CRITICAL READING SCORE: Sample Unit

To the Student

Why is the sky blue? Where do babies come from? Why do bees sting? When we were young, most of us asked questions such as these. As we learn about the world and its creatures, we begin to think that science knows all the answers. Perhaps there are no mysteries left and learning about science means simply reading reports of what others have discovered. The stories in *Weird Science* will drive such thoughts from your head. Most of them raise more questions than they answer. *Weird Science* will stir your emotions and imagination with its fifteen stories from the world of science.

While you are enjoying these thought-provoking stories, you will be developing your reading skills. This book assumes that you already are a fairly good reader. *Weird Science* is for students who want to read faster and to increase their understanding of what they read. If you complete all fifteen units—reading the stories and completing the exercises—you will surely increase your reading speed and improve your comprehension.

Group One

THE MYSTERIOUS LIFE OF TWINS

Identical twins look like each other. Many of them like to dress like each other. Often they have the same habits and the same likes and dislikes. Do they learn from each other? Are they connected in some mysterious way? Or are all their similarities due to their genes? Scientists have been exploring the relationship between twins for many years. They have studied pairs of twins who grew up together, and other pairs who were separated at birth. They have come up with some remarkable findings.

Jim Lewis was an identical twin. But he hadn't seen his brother since birth. The two boys were adopted by different families. They knew nothing about each other. Yet when they were brought together in 1979 after thirty-nine years, something spooky seemed to be going on. For one thing, both boys had been named James. Both went by the nickname "Jim." As children, they both had a pet dog named Toy.

But that was only the beginning. Each Jim had married a woman named Linda. Each then had a son. One named his son James Alan. The other named his son James Allen. Later, both Jims got divorced. Each had remarried—and in both cases, the second wife's name was Betty! Each Jim drove the same kind of blue car. Each had the same favorite drink. Each bit his nails, liked woodworking, and took vacations to the very same beach in Florida!

Could all of this be coincidence? Or do twins share a special connection? Scientists have long known that identical twins have the same genes. But no one believed there was a gene that tells you what kind of car to buy. So what made the "Jim" twins live such similar lives?

In the past, people thought twins were alike simply because they grew up together. They saw the same people. They learned to like the same things. But that is not the case with the "Jim" twins. They did not grow up together. They knew nothing about each other when they bought cars, named their sons, and picked out beaches.

In the 1980s, a man named Thomas J. Bouchard, Jr., took a closer look at twins. He found other sets of identical twins who had lived apart since birth. Among them were Daphne Goodship and Barbara Hebert. Like the "Jim" twins, these women had not seen each other for thirty-nine years. Bouchard arranged for them to meet in London, England. At that meeting, Daphne and Barbara showed up wearing the same kind of clothes! Both had chosen a light brown dress and brown velvet jacket.

As the two women compared notes, they found they were alike in many ways. Both had the weird habit of pushing up their noses. Both had met their husbands at local dances when they were sixteen. Each of them had given birth to two sons, then a daughter. Strangest of all, each had fallen down the stairs at the age of fifteen. These accidents had left both twins with weak ankles.

Then there was Jack Yufe and Oskar Stöhr. Bouchard brought them together when they were forty-seven years old. It turned out that both men had short, clipped mustaches. Both wore rectangular wire-rimmed glasses. And

Not all twins are identical twins. Fraternal twins develop from different eggs. They have no more in common with each other than other sisters and brothers do, except a common birthday. But identical twins come from the same egg. They have exactly the same genetic make-up.

both showed up at their first meeting wearing the same kind of fancy blue shirt. Jack and Oskar soon found more "coincidences." They walked with the same kind of swinging steps. They shared the habit of keeping extra rubber bands around their wrists. Both of them read magazines from back to front. They both even had the odd habit of flushing a toilet before using it!

The twins in Bouchard's study were more alike than anyone would have guessed. None of them had been in touch with his or her twin growing up. So what led them to make so many of the same choices in life? Some people think twins can communicate with each other in mysterious ways. Ron and Rod Fuller are identical twins from Dallas, Texas. They say each can tell when the other one is in trouble. Explains Rod, "There is a certain bond that we have for one another that I think all twins have."

Other twins agree. Andreina and Andreini McPherson grew up in Chino Hills, California. They say they, too, can each tell how the other is feeling. In fact, they claim, they can feel each other's pain. When one of them is hurt, the other one can feel the injury.

If that is true, then maybe twins raised apart also can communicate in special ways. Did the twins from Bouchard's study send each other messages for years without knowing it? Perhaps. But it may be that the answer lies in the genes, after all. In 1988, Dr. David Teplica began to study twins. He took pictures of six thousand pairs of identical twins. He found some amazing things. These twins had freckles in the same spots. They got gray hairs at the same time and in the same places on their heads. Their faces got the same wrinkles. They even got pimples on their noses on exactly the same day! To Dr. Teplica, there was just one way to explain all this. Genes had to be controlling these events.

It's hard to believe we are born with genes that control when and where we get pimples. But that may be the case. Thomas Bouchard says his work also points to the power of genes. He believes genes explain many of the "coincidences" among the twins he studied. So who knows? Maybe there really is a gene that tells us what kind of car to buy.

If you have been timed while reading this selection, enter your reading time below. Then turn to the Words-per-Minute table on page 110 and look up your reading speed (words per minute). Enter your reading speed on the graph on page 113.

READING TIME: Unit 1

__________ : __________

Minutes *Seconds*

How Well Did You Read?

- *Complete the four exercises that follow. The directions for each exercise will tell you how to mark your answers.*
- *When you have finished all four exercises, use the answer key on page 106 to check your work. For each right answer, put a check mark (✓) on the line beside the box. For each wrong answer, write the correct answer on the line.*
- *Follow the directions after each exercise to find your scores.*

A FINDING THE MAIN IDEA

A good main idea statement answers two questions: it tells *who* or *what* is the subject of the story, and it answers the understood question *does what?* or *is what?* Look at the three statements below. One expresses the main idea of the story you just read. Another statement is *too broad;* it is vague and doesn't tell much about the topic of the story. The third statement is *too narrow;* it tells about only one part of the story.

Match the statements with the three answer choices below by writing the letter of each answer in the box in front of the statement it goes with.

M—Main Idea B—Too Broad N—Too Narrow

____ ☐ 1. There seems to be a special connection between twins, perhaps one that is related to their shared genes.

____ ☐ 2. Jim Lewis and his twin brother lived surpisingly similar lives, from the names of their children to their choice of cars.

____ ☐ 3. The study of twins has led to some interesting discoveries.

____ Score 15 points for a correct *M* answer.
____ Score 5 points for each correct *B* or *N* answer.

____ TOTAL SCORE: Finding the Main Idea

B RECALLING FACTS

How well do you remember the facts in the story you just read? Put an *x* in the box in front of the correct answer to each of the multiple-choice questions below.

1. Scientists have long known that identical twins
   - ____ ☐ a. share the same genes.
   - ____ ☐ b. live identical lives.
   - ____ ☐ c. are always lifelong friends.

2. When twins Daphne Goodship and Barbara Hebert met for the first time in London, they both
   - ____ ☐ a. were late for the meeting.
   - ____ ☐ b. wore the same kind of clothes.
   - ____ ☐ c. had the same haircut.

3. Rod Fuller, a twin from Texas, says all twins have
   - ____ ☐ a. similar likes and dislikes.
   - ____ ☐ b. curiosity about the world.
   - ____ ☐ c. a certain bond for one another.

4. Dr. Teplica says twins' similarities are caused by
   - ____ ☐ a. the genes they share.
   - ____ ☐ b. shared experiences as they grow up.
   - ____ ☐ c. coincidence.

5. Andreina and Andreini McPherson claim that when one of them gets hurt, the other
   - ____ ☐ a. gets hurt the same way on the next day.
   - ____ ☐ b. calls her on the phone.
   - ____ ☐ c. can feel her pain.

Score 5 points for each correct answer.

____ TOTAL SCORE: Recalling Facts

C MAKING INFERENCES

When you use information from the text and your own experience to draw a conclusion that is not directly stated in the text, you are making an *inference*.

Below are five statements that may or may *not* be inferences based on the facts of the story. Write the letter *C* in the box in front of each statement that is a correct inference. Write the letter *F* in front of each faulty inference.

C—Correct Inference F—Faulty Inference

____ ☐ 1. Pairs of identical twins have the same taste in clothes.

____ ☐ 2. There is a good chance that an identical twin will look more like her twin than like her other sisters.

____ ☐ 3. Separating identical twins at birth has no effect on their closeness and friendship later in life.

____ ☐ 4. If you keep in touch with your twin, the two of you will probably begin to look alike and act alike.

____ ☐ 5. Many of the changes that take place in our bodies are caused by genes.

Score 5 points for each correct *C* or *F* answer.

____ TOTAL SCORE: Making Inferences

USING WORDS PRECISELY

Each numbered sentence below contains an underlined word or phrase from the story you have just read. Under the sentence are three definitions. One is a *synonym*, a word that means the same or almost the same thing as the underlined word: *big* and *large* are synonyms. One is an *antonym*, a word that has the opposite or nearly opposite meaning: *love* and *hate* are antonyms. One is an unrelated word; it has a completely *different* meaning than the underlined word. Match the definitions with the three answer choices by writing the letter that stands for each answer in the box in front of the definition it goes with.

S—Synonym A—Antonym D—Different

1. Could all of this be a coincidence?

___ ☐ a. chance event
___ ☐ b. planned event
___ ☐ c. joke

2. Both had the weird habit of pushing up their noses.

___ ☐ a. common
___ ☐ b. unpleasant
___ ☐ c. odd

3. Some people think twins can communicate with each other in mysterious ways.

___ ☐ a. keep things secret
___ ☐ b. make things known
___ ☐ c. quarrel

4. There is a certain bond that we have for one another that I think all twins have.

___ ☐ a. name
___ ☐ b. division
___ ☐ c. link

5. He found some amazing things.

___ ☐ a. dull and normal
___ ☐ b. incredible
___ ☐ c. similar

___ Score 3 points for a correct *S* answer.
___ Score 1 point for each correct *A* or *D* answer.

___ TOTAL SCORE: Using Words Precisely

• *Enter the total score for each exercise in the spaces below. Add the scores to find your Critical Reading Score. Then record your Critical Reading Score on the graph on page 114.*

________ Finding the Main Idea
________ Recalling Facts
________ Making Inferences
________ Using Words Precisely

________ CRITICAL READING SCORE: Unit 1

A major sign of life in a being is that it reacts to its surroundings. If it senses danger, it runs away or defends itself. It responds to food and light. It does what it can to keep itself alive. Some scientists find these characteristics in Earth itself. For example, over the ages, it has developed a protective blanket of atmosphere. Earth's temperature has risen and fallen in response to events on its surface. Dr. James Lovelock speaks for a growing number of researchers who see Earth as a living being, one they call Gaia *(guy-ah).*

IS THE EARTH ALIVE?

Imagine drilling a hole eight miles deep. That's what Arthur Conan Doyle described in his short story titled "When the World Screamed." In Doyle's story, drilling that hole turned out to be a bad idea. As the hole got deeper and deeper, the earth began to howl in pain.

Doyle's story was pure science fiction, of course. But some scientists think his image comes pretty close to the truth. These people say that our planet really is alive. It may not actually scream in pain. But it can and it does react to what we humans do.

The concept of the earth being alive may sound crazy. Most of us think of our planet as a kind of giant rock spinning through space. It is true that living creatures swarm all over this rock. But the rock itself is not alive. Or is it?

Dr. James Lovelock says it is. Lovelock is a British scientist. He calls his belief *Gaia* [pronounced guy-ah]. That means "Mother Earth" in Greek. Lovelock has written two books to explain his position. In 1979, he wrote *Gaia*. Nine years later, he wrote *The Ages of Gaia*. "You may find it hard to swallow," Lovelock said, ". . . that anything as large and apparently [dead] as the Earth is alive." Yet that's how Lovelock sees it. And he's not the first one to look at things this way. A German scientist named Gustav Fechner (1801–1887) thought everything was alive.

Fechner believed that all planets have a life of their own. In fact, he claimed, a planet is a higher form of life than you and I. As proof, Fechner noted that the earth doesn't have arms and legs. Why? According to Fechner, the earth doesn't need them. The planet Earth already has everything it desires. Human beings, on the other hand, are not born with everything they need. They must find ways to feed and shelter themselves. So they have had to develop arms and legs in order to do that.

Fechner's weird view didn't catch on during his lifetime. Other scientists simply ignored him. They went on thinking of the earth as a mixture of lava, rocks, water, soil, and plants. To be sure, these scientists said, the earth is a wonderful place. But it is not "alive" in any true sense of the word.

Then along came Lovelock. His bold views caught many people's attention. Even some scientists became interested in Gaia. Lovelock says Gaia is based on one key principle. It is this: Living things and the earth have a direct effect on each other. At first, that might not sound like a shocking idea. After all, it is clear that the earth affects life. There is no argument here. People who live in the cold mountains do things one way. Those who live in the warm tropics do things another way. Those who live in a desert do things a third way. So the conditions offered by the earth do indeed affect how we live.

But Lovelock believes the reverse is true, as well. He says that life affects the earth. To show this, he built a simple model of the world. He called it Daisyworld. The main form of life in this model world is black and white daisies. The daisies grow when it is warm and die when it is cold. But if it gets too hot or too cold, the daisies can fight back. They can get the earth to change its temperature. If the sunlight is weak, more black daisies will grow. Their black petals absorb the sunlight. This tends to warm the earth. If the sunlight is strong, more white daisies will grow. Their white petals then reflect the sunlight, which will cool the earth.

In the real world, says Lovelock, the same thing happens. Humans and other forms of life constantly cause the earth to react to what they do. Followers of Gaia believe that some of these reactions have been pretty strong. They say the earth has changed its temperature. They say it has changed the level of salt in the oceans. They even say it has moved continents around.

That does not mean humans have all the power. Lovelock notes that Mother Earth is one tough old lady. She can take a lot of abuse. After all, during her long history the earth has lived through ice ages, earthquakes, and volcanoes. The earth has even survived direct hits from meteors. It is not likely to experience anything worse. In light of what this planet has already endured, Lovelock says, a nuclear war would be "as trivial as a summer breeze."

Does this mean it doesn't matter if we blow ourselves up? That's right. Gaia followers say that if this happened, the earth itself would go right on living. And sooner or later, some other form of life would take our place. Lovelock even thinks he knows what that life form would be—whales! He says whales have brain power far beyond what we have imagined.

Many people still think Lovelock and his followers are loony. Still, Gaia has a magical ring to it. The idea is catching on. There have been dozens of articles written about it. There have been Gaia lectures. There have been Gaia films. A 1984 book on Gaia sold more than 175,000 copies. No one yet claims to have heard the earth crying out like it did in Doyle's story. But maybe, just maybe, we're not listening hard enough.

If you have been timed while reading this selection, enter your reading time below. Then turn to the Words-per-Minute table on page 110 and look up your reading speed (words per minute). Enter your reading speed on the graph on page 113.

READING TIME: Unit 2

__________ : __________

Minutes *Seconds*

How Well Did You Read?

- *Complete the four exercises that follow. The directions for each exercise will tell you how to mark your answers.*
- *When you have finished all four exercises, use the answer key on page 106 to check your work. For each right answer, put a check mark (✓) on the line beside the box. For each wrong answer, write the correct answer on the line.*
- *Follow the directions after each exercise to find your scores.*

A FINDING THE MAIN IDEA

A good main idea statement answers two questions: it tells *who* or *what* is the subject of the story, and it answers the understood question *does what?* or *is what?* Look at the three statements below. One expresses the main idea of the story you just read. Another statement is *too broad*; it is vague and doesn't tell much about the topic of the story. The third statement is *too narrow*; it tells about only one part of the story.

Match the statements with the three answer choices below by writing the letter of each answer in the box in front of the statement it goes with.

M—Main Idea B—Too Broad N—Too Narrow

___ ☐ 1. Not all scientists agree with Dr. James Lovelock on the nature of the planet Earth.

___ ☐ 2. Dr. James Lovelock calls his concept Gaia, meaning "Mother Earth" in Greek.

___ ☐ 3. The Gaia concept says that Earth reacts like a living being to changing conditions.

___ Score 15 points for a correct *M* answer.
___ Score 5 points for each correct *B* or *N* answer.

___ TOTAL SCORE: Finding the Main Idea

B RECALLING FACTS

How well do you remember the facts in the story you just read? Put an *x* in the box in front of the correct answer to each of the multiple-choice questions below.

1. Dr. James Lovelock is
   - ____ ☐ a. a writer of science fiction.
   - ____ ☐ b. an American scientist.
   - ____ ☐ c. a British scientist.

2. The German scientist Gustav Fechner believed that
   - ____ ☐ a. digging a deep hole can make Earth scream.
   - ____ ☐ b. planets are a higher life form than humans.
   - ____ ☐ c. anything as large as Earth is dead.

3. The principle underlying Gaia is that living things
   - ____ ☐ a. depend on Earth for their survival.
   - ____ ☐ b. must find ways to feed and shelter themselves.
   - ____ ☐ c. and Earth have a direct effect on each other.

4. If more black daisies grow on Earth, their petals absorb sunlight and
   - ____ ☐ a. Earth gets warmer.
   - ____ ☐ b. Earth gets cooler.
   - ____ ☐ c. there is no effect on Earth's temperature.

5. Followers of Gaia say that activity on Earth has
   - ____ ☐ a. brought on direct hits from meteors.
   - ____ ☐ b. caused changes in the level of salt in oceans.
   - ____ ☐ c. given humans power over the planet.

Score 5 points for each correct answer.

____ TOTAL SCORE: Recalling Facts

C MAKING INFERENCES

When you use information from the text and your own experience to draw a conclusion that is not directly stated in the text, you are making an *inference*.

Below are five statements that may or may not be inferences based on the facts of the story. Write the letter *C* in the box in front of each statement that is a correct inference. Write the letter *F* in front of each faulty inference.

C—Correct Inference **F—Faulty Inference**

____ ☐ 1. Most science fiction writers believe in the Gaia concept.

____ ☐ 2. What living things on Earth did during the last ice age may have caused Earth's temperature to rise, bringing an end to the Ice Age.

____ ☐ 3. A person who believes in God could not believe in Gaia.

____ ☐ 4. Even if Earth reacts to living creatures on its surface, it won't favor one life form over another.

____ ☐ 5. People who believe in Gaia want to change human behavior in order to save the planet from death.

Score 5 points for each correct *C* or *F* answer.

____ TOTAL SCORE: Making Inferences

D USING WORDS PRECISELY

Each numbered sentence below contains an underlined word or phrase from the story you have just read. Under the sentence are three definitions. One is a *synonym*, a word that means the same or almost the same thing as the underlined word: *big* and *large* are synonyms. One is an *antonym*, a word that has the opposite or nearly opposite meaning: *love* and *hate* are antonyms. One is an unrelated word; it has a completely *different* meaning than the underlined word. Match the definitions with the three answer choices by writing the letter that stands for each answer in the box in front of the definition it goes with.

S—Synonym **A—Antonym** **D—Different**

1. But it can and does <u>react</u> to what we humans do.

___ ☐ a. give no attention
___ ☐ b. act in answer
___ ☐ c. come closer

2. It is true that living creatures <u>swarm</u> all over this rock.

___ ☐ a. crowd
___ ☐ b. jog
___ ☐ c. wait alone and unmoving

3. You may find it hard to <u>swallow</u> . . . that anything as large and apparently dead as the Earth is alive.

___ ☐ a. reject
___ ☐ b. describe
___ ☐ c. accept

4. But Lovelock believes the <u>reverse</u> is true, as well.

___ ☐ a. opposite
___ ☐ b. same
___ ☐ c. course

5. In light of what this planet has already endured, Lovelock says, a nuclear war would be "as <u>trivial</u> as a summer breeze."

___ ☐ a. unlikely
___ ☐ b. unimportant
___ ☐ c. serious

___ Score 3 points for a correct *S* answer.
___ Score 1 point for each correct *A* or *D* answer.

___ TOTAL SCORE: Using Words Precisely

• *Enter the total score for each exercise in the spaces below. Add the scores to find your Critical Reading Score. Then record your Critical Reading Score on the graph on page 114.*

______ Finding the Main Idea
______ Recalling Facts
______ Making Inferences
______ Using Words Precisely

______ CRITICAL READING SCORE: Unit 2

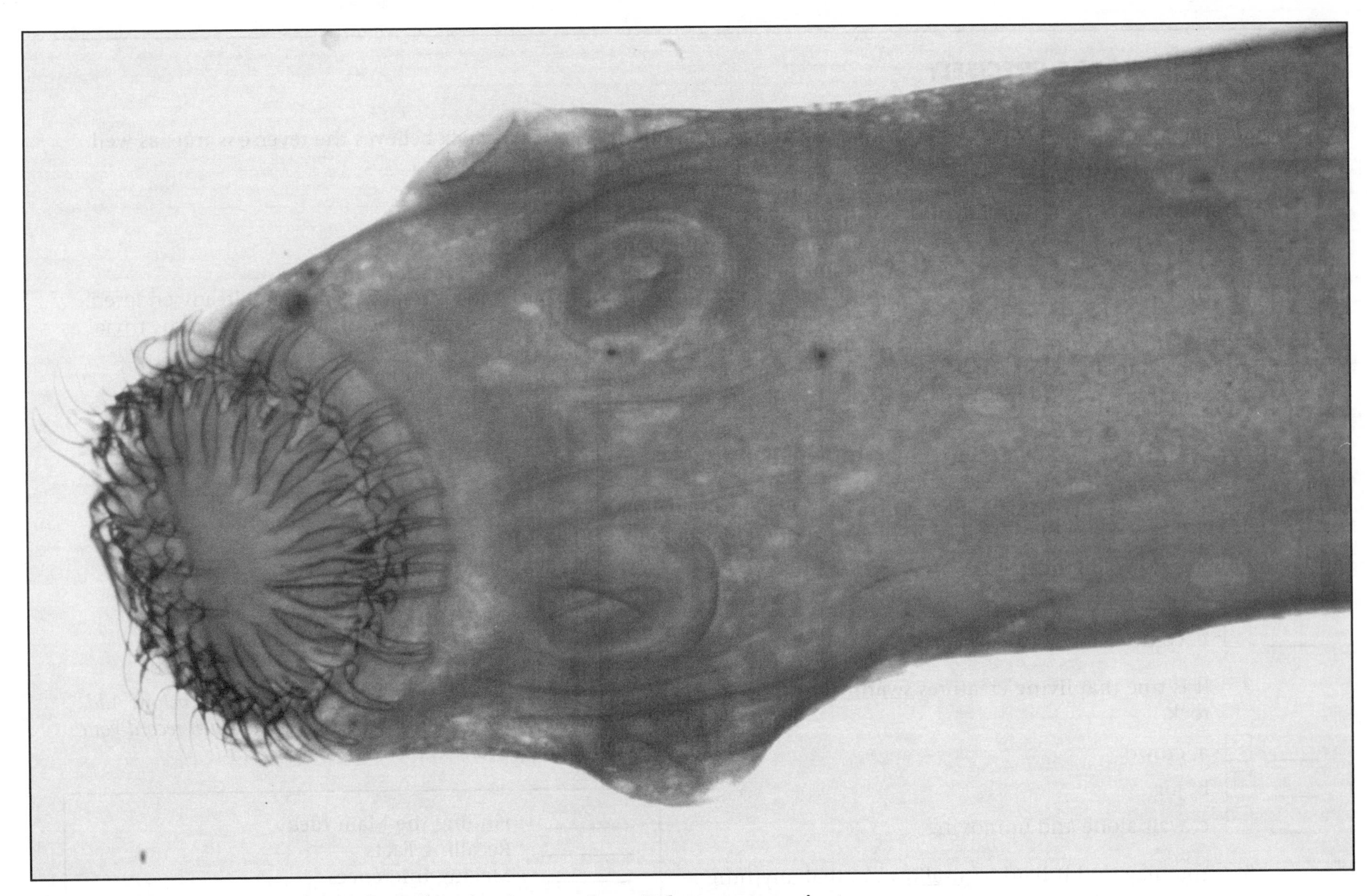

It has no legs or fins, because it usually stays in one place. It has no arms, just suckers and hooks with which it clings to the inside of your intestines. It has no mouth—it absorbs your food through its body wall. Coiled inside your intestines, it adds blocklike segments, one after another, until the string of segments reaches a length of about thirty feet. It is a tapeworm, and it is one of the least dangerous worms that live inside humans. Other parasitic worms can be **really** *unpleasant!*

WORMS, WORMS, WORMS

An old joke asks the question, "What's worse than finding a worm in your apple?" The answer is, "Finding half a worm in your apple." Actually, you have nothing to fear from swallowing ordinary worms. They may not taste very good. But they won't do you any harm. Some worms, however, are dangerous to humans. If they get inside your body, you could be in a lot of trouble!

The dangerous worms are parasites. That means they live by attaching themselves to other living creatures. They feed off the blood or body parts of "host" animals. Sometimes the host is a dog. Sometimes it is a fish. Sometimes it is a human being.

Parasitic worms start out as tiny eggs. They might be present in water, soil, or even grass. They may also be found in raw meat. The trouble begins when these eggs get inside a human body. That might happen if you drink dirty water. It could happen if you put dirty hands into your mouth. Or it might happen if you eat meat that has not been properly cooked.

After the eggs get inside a body, they pop open. Tiny worms come swimming out. The worms make their way to a particular part of the host's body. Some parasitic worms head for muscles. Others prefer the eyes. Many lodge in the intestines.

Wherever they go, the worms make themselves right at home. As they feed off their host, they grow larger and larger. Some, such as the tapeworm, can reach twenty-five or thirty feet. A worm that size would stretch from one end of a telephone pole to the other! Imagine having it curled up inside your intestines. It's not a pleasant thought.

Actually, tapeworms are not as bad as some worms. Tapeworms may be long, but they are fairly harmless. They simply make your stomach a little sore. And they may cause you to feel hungry all the time. (Tapeworms have big appetites. Much of what you eat, they take away from you.)

Other worms do much more damage. The one called "trichina" causes muscles to become stiff. Pinworms cause skin to itch. And eye worms may leave lumps the size of hen's eggs as they travel under the host's skin.

It gets worse. Consider the Guinea worm. It has caused problems in many African countries. Guinea worms begin their life in dirty water. If someone drinks that water, the microscopic worms are carried into the stomach. From there, they move down through the body. Their goal is to get to the feet. To do that, they burrow their way through muscles. They bore through bones. Sometimes they try to wiggle their way through the heart or spinal cord. If they do that, their host will die.

Even in less serious cases, Guinea worms do plenty of harm. They cause painful infections. Many people with Guinea worms are in too much pain to get out of bed.

As the Guinea worms travel, they grow. They may get to be three feet long. They become as thick as a piece of spaghetti. When they reach the feet, they dig their way outward, toward the skin. Then they settle down to wait. The next time their host steps in water, the worms break through the skin. They deposit new eggs into the water. Those eggs lie in the water, ready to be swallowed by the next victim. When that happens, the cycle begins again.

Hookworms are another kind of parasitic worm. These tiny worms can get inside a body by squirming right through the skin. They make their way into the bloodstream. From there, they are carried along into the lungs. Later, the worms slither up out of the lungs into the mouth. When the host swallows, the worms are flushed down through the stomach to the intestines.

It is here, in the intestines, that hookworms make their home. Adult hookworms have four small teeth shaped like hooks. They use those hooks to clamp onto the wall of the intestines. They can stay hooked on like that for years, sucking blood from their host.

People with hookworms are constantly being robbed of blood. That makes them tired. Their skin turns pale. If their blood supply falls too low, their brains can be affected. Their senses may become dull. They may lose interest in things around them. One or two hookworms won't cause such severe problems. But sometimes hundreds of worms get into one body. In Indonesia recently, one hundred-eighty hookworms were pulled from one man's body. At times, over five hundred hookworms have been found in one person.

And there is the roundworm. It, too, loves the human intestine. Usually it follows the same route as the hookworm. It goes from the blood to the lungs to the mouth. From there, it goes down to the intestine. But sometimes roundworms take a wrong turn. Sometimes they end up in other spots. That can mean big trouble for the host. If a roundworm winds up in the brain, it spells death for the host. It creates infections that cannot be controlled. That happened to a three-year-old Michigan girl in 1981. There was nothing doctors could do for little Roberta Hansen. She died a year after roundworms entered her brain.

Roberta Hansen's case was very unusual. Most Americans never have to deal with parasitic worms. Countries with high health standards have few problems with these creatures. Still, you might think twice before gulping down any dirty water or eating poorly cooked meat.

If you have been timed while reading this selection, enter your reading time below. Then turn to the Words-per-Minute table on page 110 and look up your reading speed (words per minute). Enter your reading speed on the graph on page 113.

READING TIME: Unit 3

__________ : __________

Minutes *Seconds*

How Well Did You Read?

- *Complete the four exercises that follow. The directions for each exercise will tell you how to mark your answers.*
- *When you have finished all four exercises, use the answer key on page 106 to check your work. For each right answer, put a check mark (✓) on the line beside the box. For each wrong answer, write the correct answer on the line.*
- *Follow the directions after each exercise to find your scores.*

FINDING THE MAIN IDEA

A good main idea statement answers two questions: it tells *who* or *what* is the subject of the story, and it answers the understood question *does what?* or *is what?* Look at the three statements below. One expresses the main idea of the story you just read. Another statement is *too broad*; it is vague and doesn't tell much about the topic of the story. The third statement is *too narrow*; it tells about only one part of the story.

Match the statements with the three answer choices below by writing the letter of each answer in the box in front of the statement it goes with.

M—Main Idea B—Too Broad N—Too Narrow

____ ☐ 1. Human beings often serve as hosts to parasites such as worms.

____ ☐ 2. Certain worms that live inside humans and feed off them can cause serious health problems.

____ ☐ 3. Some worms that live inside humans, such as the tapeworm, can grow to twenty-five or thirty feet in length.

____ Score 15 points for a correct *M* answer.
____ Score 5 points for each correct *B* or *N* answer.

____ TOTAL SCORE: Finding the Main Idea

B RECALLING FACTS

How well do you remember the facts in the story you just read? Put an *x* in the box in front of the correct answer to each of the multiple-choice questions below.

1. Parasites are
   ___ ☐ a. worms that live inside human beings.
   ___ ☐ b. living things that live by attaching themselves to other living creatures.
   ___ ☐ c. tiny eggs that are dangerous to human beings.

2. Parasitic worm eggs can get inside humans when
   ___ ☐ a. the humans get overheated.
   ___ ☐ b. the humans eat uncooked meat.
   ___ ☐ c. the adult worms are about half an inch long.

3. Once inside a human host, parasitic worms usually
   ___ ☐ a. stay in the intestine until the worms die.
   ___ ☐ b. travel to a particular part of the host's body.
   ___ ☐ c. cause intense pain and the host's death.

4. Guinea worms travel down to the host's feet to
   ___ ☐ a. live out their lives there.
   ___ ☐ b. cause egg-shaped lumps on the toes.
   ___ ☐ c. break through the skin and lay eggs outside.

5. Hookworms live off their host by taking the host's
   ___ ☐ a. blood.
   ___ ☐ b. food.
   ___ ☐ c. intestines.

Score 5 points for each correct answer.

___ TOTAL SCORE: Recalling Facts

C MAKING INFERENCES

When you use information from the text and your own experience to draw a conclusion that is not directly stated in the text, you are making an *inference*.

Below are five statements that may or may *not* be inferences based on the facts of the story. Write the letter *C* in the box in front of each statement that is a correct inference. Write the letter *F* in front of each faulty inference.

C—Correct Inference **F—Faulty Inference**

___ ☐ 1. Ordinary worms are not parasitic.

___ ☐ 2. You can avoid eating meat that contains the eggs of parasitic worms by examining the meat carefully.

___ ☐ 3. If clean drinking water were available in Africa, far fewer people would suffer from Guinea worms.

___ ☐ 4. After a Guinea worm has dug its way out of the host's body, the host will get well without medical treatment.

___ ☐ 5. People with hookworms or roundworms can feel the worms traveling through their bodies.

Score 5 points for each correct *C* or *F* answer.

___ TOTAL SCORE: Making Inferences

USING WORDS PRECISELY

Each numbered sentence below contains an underlined word or phrase from the story you have just read. Under the sentence are three definitions. One is a *synonym*, a word that means the same or almost the same thing as the underlined word: *big* and *large* are synonyms. One is an *antonym*, a word that has the opposite or nearly opposite meaning: *love* and *hate* are antonyms. One is an unrelated word; it has a completely *different* meaning than the underlined word. Match the definitions with the three answer choices by writing the letter that stands for each answer in the box in front of the definition it goes with.

S—Synonym A—Antonym D—Different

1. If someone drinks that water, the microscopic worms are carried into the stomach.

___ ☐ a. bulky
___ ☐ b. very tiny
___ ☐ c. dangerous

2. To do that, they burrow their way through muscles.

___ ☐ a. dig
___ ☐ b. announce
___ ☐ c. leak

3. They deposit new eggs into the water.

___ ☐ a. stir up
___ ☐ b. put down
___ ☐ c. swallow

4. Later, the worms slither up out of the lungs into the mouth.

___ ☐ a. move in a mechanical way
___ ☐ b. hurry
___ ☐ c. move like a snake

5. If their blood supply falls too low, their brains can be affected.

___ ☐ a. measured
___ ☐ b. ignored
___ ☐ c. influenced

___ Score 3 points for a correct *S* answer.
___ Score 1 point for each correct *A* or *D* answer.

___ TOTAL SCORE: Using Words Precisely

- *Enter the total score for each exercise in the spaces below. Add the scores to find your Critical Reading Score. Then record your Critical Reading Score on the graph on page 114.*

________	Finding the Main Idea
________	Recalling Facts
________	Making Inferences
________	Using Words Precisely
________	CRITICAL READING SCORE: Unit 3

Firefighters battle a huge blaze that has already burned over 3000 acres in the American West. Wind whips the flames across the dry bushes and weeds. Anyone living nearby can only hope for heavy rains to help put out the fire. As deadly and destructive as this blaze is, it cannot compare to the fury of a fire storm. A true fire storm is so intense and violent that fire fighters would have no effect on it. Luckily, this phenomenon occurs only under special conditions. It destroys everything and everyone nearby, and it burns until nothing is left to burn.

FIRE STORMS

Legend says that it all began with a cow. On October 8, 1871, Mrs. O'Leary's cow knocked over a lantern in a barn. A fire broke out. Soon most of Chicago was going up in flames. The damage was immense. Fire destroyed more than two thousand acres of the city. Over one hundred thousand people lost their homes. About three hundred people died in the inferno. The Great Chicago Fire of 1871 shocked the nation. It became the most famous fire in American history.

And, yet, on the same day, there was a fire in Wisconsin that was even worse. People did not hear about this second fire right away. The Great Chicago Fire was the hot news story of the day. Besides, the second fire took place in the small lumber town of Peshtigo. It took several days for word from this town to reach the outside world.

Peshtigo had seen plenty of fires before. Dense woods surrounded the town. Brush fires often broke out. The people of Peshtigo knew how to deal with them. This fire, though, was impossible to control. About 9:30 P.M., someone saw a dull red glow in the distance. That was followed by a low rumbling sound. Everyone in town knew exactly what that meant. The men jumped into action. The women got the children out of bed and dressed them. By ten o'clock, the woods had turned bright crimson as flames leaped from tree to tree. Sparks flew everywhere. Soon the blaze reached the town itself. The wooden sidewalks caught fire. Sawdust used in the streets to keep the dust down also burst into flames. The angry blaze engulfed one building after another.

There was no hope of stopping the fire. The people just tried to save themselves. Some sought shelter in large buildings. But as the buildings went up in flames, most of these people burned to death. Others drowned after leaping into the river. Three people jumped into a large water tank at a sawmill. But even they did not survive. The fire turned the water so hot that everyone in the tank died.

The Peshtigo fire destroyed every building in town. About eight hundred people died. That was five hundred more people than the Great Chicago Fire killed. In terms of lost lives, then, the Peshtigo fire was much worse than the one in Chicago.

The fame of the Chicago fire is well earned. It was, after all, a truly massive blaze. But it was a regular fire. The one in Peshtigo, on the other hand, was a rare kind of fire. It was actually a "fire storm." People who survived it talked of winds that were "tornado-like." They said balls of fire seemed to jump out of nowhere. These balls appeared and disappeared like lightning.

What is the difference between a normal fire and a fire storm? A normal fire is largely controlled by the weather. High winds can fan the flames. In fact, strong, gusty winds did help to spread the Chicago fire. Similarly, a heavy rain can douse a normal fire. For example, rain often checks forest fires. A fire storm, on the other hand, *creates* its own weather. It makes its *own* wind and rain. A fire storm can make rain fall and lightning flash even on a sunny day. It can create small tornadoes, or whirls, filled with fire and deadly gases. These little weather systems grow inside a plume of smoke that rises high above the ground.

Fire storms are rare. The conditions have to be just right to create one. First, the fire must be really hot. That means having lots of fuel such as dry wood, sawdust, twigs, and brush. Second, the winds in the area must be weak. A strong wind would blow the rising smoke across the land and keep a plume from developing. Third, the air must be fairly warm. Warm air forms currents that rise into the upper atmosphere. Cold air sinks. Cold air would press down on the plume and keep it from growing.

If the conditions are met, watch out. A billowing plume develops. It carries heat, smoke, ash, and gases higher and higher. Within this plume, the wind whips around at very high speeds. This wind turns into small but deadly tornadoes. The tornadoes can be as high as four hundred feet and as wide as fifty feet. They travel at speeds of just six or seven miles an hour. But it's hard to tell where they'll go next.

As the plume rises, moisture in the air starts to condense on the ash and smoke particles. This creates a cloud that looks like a towering black storm cloud. A 1993 fire storm in Santa Barbara, California, created such a cloud. It reached thirty-eight thousand feet. That's almost two miles higher than Mount Everest!

As the cloud grows, more and more moisture condenses on the ash and smoke particles. Soon rain starts to fall. In that way, a fire storm creates its own rainfall. The Santa Barbara fire storm produced lightning and almost half an inch of rain. But such rain rarely puts out the fire. One reason is that the plume doesn't stay perfectly straight. The upper part, where the rain forms, drifts slowly away from the source of the fire. So the rain doesn't fall on the fire itself. As a result, most fire storms don't put themselves out. They die only when their fuel supply runs out.

If you have been timed while reading this selection, enter your reading time below. Then turn to the Words-per-Minute table on page 110 and look up your reading speed (words per minute). Enter your reading speed on the graph on page 113.

READING TIME: Unit 4

__________ : __________

Minutes *Seconds*

How Well Did You Read?

- *Complete the four exercises that follow. The directions for each exercise will tell you how to mark your answers.*
- *When you have finished all four exercises, use the answer key on page 106 to check your work. For each right answer, put a check mark (✓) on the line beside the box. For each wrong answer, write the correct answer on the line.*
- *Follow the directions after each exercise to find your scores.*

A FINDING THE MAIN IDEA

A good main idea statement answers two questions: it tells *who* or *what* is the subject of the story, and it answers the understood question *does what?* or *is what?* Look at the three statements below. One expresses the main idea of the story you just read. Another statement is *too broad;* it is vague and doesn't tell much about the topic of the story. The third statement is *too narrow;* it tells about only one part of the story.

Match the statements with the three answer choices below by writing the letter of each answer in the box in front of the statement it goes with.

M—Main Idea B—Too Broad N—Too Narrow

____ ☐ 1. Fire storms such as the one in Peshtigo in 1871 are a special kind of fire.

____ ☐ 2. A fire storm destroyed the small lumber town of Peshtigo in October 1871, on the same day as the Great Chicago Fire.

____ ☐ 3. Fire storms are more dangerous than regular fires because they make their own weather and cannot be controlled.

____ Score 15 points for a correct *M* answer.
____ Score 5 points for each correct *B* or *N* answer.

____ TOTAL SCORE: Finding the Main Idea

B RECALLING FACTS

How well do you remember the facts in the story you just read? Put an *x* in the box in front of the correct answer to each of the multiple-choice questions below.

1. Fires often started near Peshtigo because the
   - ____ ☐ a. town was near Chicago.
   - ____ ☐ b. town was surrounded by dense woods.
   - ____ ☐ c. people of the town were careless.

2. In a regular fire, heavy winds can
   - ____ ☐ a. put out the fire.
   - ____ ☐ b. start the fire.
   - ____ ☐ c. spread the fire.

3. The plume of a fire storm can reach heights of
   - ____ ☐ a. about forty thousand feet.
   - ____ ☐ b. about two thousand feet.
   - ____ ☐ c. about four hundred feet.

4. One of the conditions that must be met before a fire storm can develop is
   - ____ ☐ a. fairly warm air.
   - ____ ☐ b. strong winds.
   - ____ ☐ c. a cold weather front.

5. As the plume rises, moisture in the air
   - ____ ☐ a. turns to ice.
   - ____ ☐ b. disappears in the heat from the fire.
   - ____ ☐ c. condenses on the ash and smoke particles.

Score 5 points for each correct answer.

____ TOTAL SCORE: Recalling Facts

C MAKING INFERENCES

When you use information from the text and your own experience to draw a conclusion that is not directly stated in the text, you are making an *inference*.

Below are five statements that may or may *not* be inferences based on the facts of the story. Write the letter *C* in the box in front of each statement that is a correct inference. Write the letter *F* in front of each faulty inference.

C—Correct Inference F—Faulty Inference

____ ☐ 1. The citizens of Peshtigo knew the roles they were to play whenever fire broke out.

____ ☐ 2. News did not travel as fast in 1871 as it does today.

____ ☐ 3. It is likely that you will see several fire storms in your lifetime.

____ ☐ 4. Most tornadoes are caused by fire storms.

____ ☐ 5. In a desert area without brush or dry wood, a fire storm would probably not develop.

Score 5 points for each correct *C* or *F* answer.

____ TOTAL SCORE: Making Inferences

D USING WORDS PRECISELY

Each numbered sentence below contains an underlined word or phrase from the story you have just read. Under the sentence are three definitions. One is a *synonym*, a word that means the same or almost the same thing as the underlined word: *big* and *large* are synonyms. One is an *antonym*, a word that has the opposite or nearly opposite meaning: *love* and *hate* are antonyms. One is an unrelated word; it has a completely *different* meaning than the underlined word. Match the definitions with the three answer choices by writing the letter that stands for each answer in the box in front of the definition it goes with.

S—Synonym A—Antonym D—Different

1. It was after all, a truly massive blaze.

___ ☐ a. huge
___ ☐ b. terrifying
___ ☐ c. tiny

2. People who survived it talked of winds that were "tornado-like."

___ ☐ a. died in
___ ☐ b. remembered
___ ☐ c. lived through

3. For example, rain often checks forest fires.

___ ☐ a. stops
___ ☐ b. encourages
___ ☐ c. goes along with

4. This creates a cloud that looks like a towering black storm cloud.

___ ☐ a. short
___ ☐ b. tall
___ ☐ c. fast-moving

5. The Great Chicago Fire of 1871 shocked the nation.

___ ☐ a. disturbed
___ ☐ b. fooled
___ ☐ c. bored

___ Score 3 points for a correct *S* answer.
___ Score 1 point for each correct *A* or *D* answer.

___ TOTAL SCORE: Using Words Precisely

- *Enter the total score for each exercise in the spaces below. Add the scores to find your Critical Reading Score. Then record your Critical Reading Score on the graph on page 114.*

______	Finding the Main Idea
______	Recalling Facts
______	Making Inferences
______	Using Words Precisely
______	CRITICAL READING SCORE: Unit 4

What is this woman doing? It looks like she's simply walking around a field with a stick in her hands. But look carefully at the stick. It is forked. People known as dowsers *use such sticks as a kind of sensor. With a dowsing rod, they can find things hidden underground. Dowsers have been able to find metal objects (including pipes and land mines), ancient relics, and—most importantly—water. For over three hundred years, people have been using this technique. Today, scientists are trying to find out if it really works and, if so, how.*

DOWSING: FACT OR FICTION?

Ray Burbank was having trouble finding an underground water pipe. So he asked a friend named Henry Gross for help. Gross took a Y-shaped twig and held it in his hands. Then he walked back and forth over the ground. "The pipe's right here," Gross said at last, marking the spot with a wooden stake.

Meanwhile, the water company had sent its own men to find the pipe. When the men saw Gross with the twig, they broke out laughing. Still, even though they used fancy machines, they couldn't find the pipe. The next day, Ray Burbank dug up the spot Gross had marked. Sure enough, there was the water pipe.

Gross found the pipe by using the age-old art of dowsing. Dowsers claim they can find water and other hidden things under the earth. They simply walk over the ground while holding a forked stick or rod. Suddenly, they say, the stick or rod will tremble in the dowser's hands. It will point down toward what is hidden below the ground. When asked what makes the stick move, many dowsers shrug. "I don't know how it works," they say. "It just does."

Henry Gross is not the only dowser to amaze his neighbors. An old Vermont farmer named Milford Preston was famous for picking the best place to drill for water. One day a friend challenged Preston to a test. The friend dumped five piles of sand behind his barn. He told Preston he had hidden a quarter in one of the piles. In truth, the friend was trying to trick Preston. He had actually hidden quarters in two different piles.

Preston picked up a forked stick and went to work. He stopped over the second pile. He could tell a quarter was buried there. But to be sure, Preston checked the other piles. He knew right away that there was another quarter in the fifth pile. "You're not as tricky as you thought you were!" Preston smirked.

Dowsing goes back at least to the 16th century. That's when the first written account of it appeared in Germany. In those days, dowsing was used to find precious metals. The practice spread throughout Europe and, later, the United States. People in Asia and Africa also began to practice dowsing. Over time, dowsers have expanded their claims. Today they still say they can find water, pipelines, and metals. But they also say they can locate buried treasure. They claim they can find ancient relics, land mines, and dead bodies. Some dowsers even insist they can find objects just by swinging a chain over a map. The chain, they say, will pull their hand toward the right spot.

Some of today's dowsers have a pretty good record of success. One of the very best dowsers is Hans Schröter. Schröter has spent a lot of time in Sri Lanka. He has picked sites for hundreds of wells there. In fact, he has chosen 691 spots. Only twenty-seven of these have failed to yield water.

Still, the question remains: How do dowsers do it? What could make a stick suddenly bend down toward something far underground? Is there some force in nature at work? Some people think there is. They believe each hidden object must send out some kind of mysterious wave. Water, too, must send out waves. Dowsers, then, would be people sensitive enough to pick up these waves.

Few scientists believe in such unseen waves. Some say that dowsers' success stories are just a matter of luck. Others have a different theory. It's not the stick that helps a dowser, they say. It's the dowser's own knowledge of the land. Most dowsers are not geologists. They have no formal training in earth science. Still, they often know the land they are walking very well. So they might pick up clues without even realizing it. They might see that underground water changes the look of the soil in a certain area. The shape of the ground might offer hints. So, too, might the presence of certain plants or grasses. Geologist Jay Lehr says that experienced dowsers are often experts in picking up such clues. He says dowsers always "have an understanding, whether they're aware of it or not, of various surface clues."

Still, a few experts have decided that dowsing is for real. One is the German scientist Hans-Dieter Betz. In 1995, Betz wrote a report on dowsing. In it, he declared that good dowsers can indeed detect water below the ground. Betz is respected in his field. His report has caused other scientists to take a second look at dowsing. So far, though, most are not convinced.

So that puts us back where we started. Is dowsing fact or fiction? Tests have shown that it does work. But all of these tests, including ones done by Betz, have been challenged. Critics of dowsing say that every test has been flawed in one way or another. So dowsing remains an open question. Most scientists still reject it. But millions of people around the world practice it. Can they all be wrong?

If you have been timed while reading this selection, enter your reading time below. Then turn to the Words-per-Minute table on page 110 and look up your reading speed (words per minute). Enter your reading speed on the graph on page 113.

READING TIME: Unit 5

_____________ : _____________

Minutes *Seconds*

How Well Did You Read?

- *Complete the four exercises that follow. The directions for each exercise will tell you how to mark your answers.*
- *When you have finished all four exercises, use the answer key on page 106 to check your work. For each right answer, put a check mark (✓) on the line beside the box. For each wrong answer, write the correct answer on the line.*
- *Follow the directions after each exercise to find your scores.*

A FINDING THE MAIN IDEA

A good main idea statement answers two questions: it tells *who* or *what* is the subject of the story, and it answers the understood question *does what?* or *is what?* Look at the three statements below. One expresses the main idea of the story you just read. Another statement is *too broad*; it is vague and doesn't tell much about the topic of the story. The third statement is *too narrow*; it tells about only one part of the story.

Match the statements with the three answer choices below by writing the letter of each answer in the box in front of the statement it goes with.

M—Main Idea B—Too Broad N—Too Narrow

____ ☐ 1. Dowsers use a forked stick or rod that trembles when it is placed over a buried object.

____ ☐ 2. Dowsing is a good way to find hidden objects.

____ ☐ 3. The dowsing method of finding buried objects seems to be a combination of art, skill, and luck.

____ Score 15 points for a correct *M* answer.
____ Score 5 points for each correct *B* or *N* answer.

____ TOTAL SCORE: Finding the Main Idea

B RECALLING FACTS

How well do you remember the facts in the story you just read? Put an *x* in the box in front of the correct answer to each of the multiple-choice questions below.

1. When the water company men saw the dowser whom Ray Burbank had hired, they
   - ____ ☐ a. made him leave the land.
   - ____ ☐ b. asked him for help.
   - ____ ☐ c. laughed at him.

2. Dowsers find water and hidden objects by
   - ____ ☐ a. walking over ground with a forked stick.
   - ____ ☐ b. listening to the sound a forked stick makes.
   - ____ ☐ c. closing their eyes and thinking hard.

3. The first written account of dowsing came from
   - ____ ☐ a. Africa.
   - ____ ☐ b. Mexico.
   - ____ ☐ c. Germany.

4. Some dowsers say that when they want to find something, a chain swung over a map can
   - ____ ☐ a. make them dizzy.
   - ____ ☐ b. pull their hand toward the right spot.
   - ____ ☐ c. help them think better.

5. When the forked stick is held over a buried object, the stick
   - ____ ☐ a. bends down toward the ground.
   - ____ ☐ b. jumps out of the dowser's hand.
   - ____ ☐ c. becomes hot in the dowser's hand.

Score 5 points for each correct answer.

____ TOTAL SCORE: Recalling Facts

C MAKING INFERENCES

When you use information from the text and your own experience to draw a conclusion that is not directly stated in the text, you are making an *inference*.

Below are five statements that may or may *not* be inferences based on the facts of the story. Write the letter *C* in the box in front of each statement that is a correct inference. Write the letter *F* in front of each faulty inference.

C—Correct Inference F—Faulty Inference

____ ☐ 1. One of the advantages of dowsing is that it can be done with cheap tools.

____ ☐ 2. Only men can become dowsers.

____ ☐ 3. The need for wells in Sri Lanka is great.

____ ☐ 4. If scientists say they don't believe in a practice such as dowsing, everyone around the world gives it up.

____ ☐ 5. Most dowsers have studied about rocks and the earth in college.

Score 5 points for each correct *C* or *F* answer.

____ TOTAL SCORE: Making Inferences

D USING WORDS PRECISELY

Each numbered sentence below contains an underlined word or phrase from the story you have just read. Under the sentence are three definitions. One is a *synonym*, a word that means the same or almost the same thing as the underlined word: *big* and *large* are synonyms. One is an *antonym*, a word that has the opposite or nearly opposite meaning: *love* and *hate* are antonyms. One is an unrelated word; it has a completely *different* meaning than the underlined word. Match the definitions with the three answer choices by writing the letter that stands for each answer in the box in front of the definition it goes with.

S—Synonym A—Antonym D—Different

1. Only twenty-seven of these have failed to yield water.

___ ☐ a. hold back
___ ☐ b. spray
___ ☐ c. give up

2. Suddenly, they say, the stick or rod will tremble in their hands.

___ ☐ a. shake
___ ☐ b. remain steady
___ ☐ c. twist

3. In those days, dowsing was used to find precious metals.

___ ☐ a. beautiful
___ ☐ b. valuable
___ ☐ c. worthless

4. Some dowsers even insist they can find objects just by swinging a chain over a map.

___ ☐ a. deny
___ ☐ b. believe
___ ☐ c. say firmly

5. Critics of dowsing say that every test has been flawed in one way or another.

___ ☐ a. perfect
___ ☐ b. imperfect
___ ☐ c. hard to believe

___ Score 3 points for a correct *S* answer.
___ Score 1 point for each correct *A* or *D* answer.

___ TOTAL SCORE: Using Words Precisely

• *Enter the total score for each exercise in the spaces below. Add the scores to find your Critical Reading Score. Then record your Critical Reading Score on the graph on page 114.*

_______ Finding the Main Idea
_______ Recalling Facts
_______ Making Inferences
_______ Using Words Precisely

_______ CRITICAL READING SCORE: Unit 5

Group Two.

Who can forget the modified DeLorean sports car that carries Marty McFly back in time in the hit movie Back to the Future*? An even stranger vehicle carries the hero of the movie* The Time Machine *(based on an 1895 H.G. Wells novel) forward in time. The idea of time travel has had a grip on our imagination for centuries. Even the famous scientist Albert Einstein considered this possibility as he developed his theory of relativity. His work is the basis for today's scientific thinking about time travel.*

TRAVELING THROUGH TIME

Imagine being able to travel two hundred years into the future. Or think about taking a journey far back into the past. Time travel has long been a dream for many people. But is it just a dream? Until recently, everyone thought so. It was fun to ponder, scientists said, but it wasn't really possible. Now, some scientists are changing their minds. They say that maybe, just maybe, time travel is possible.

Think about time for a minute. What is it, really? You can watch the second hand on a clock move around the dial. If you watch it long enough, you'll see the minute hand move. And if you wait even longer, you'll notice that the hour hand also moves. The clock on the wall is one measure of time. But it is not the only measure.

A great scientist named Albert Einstein showed that time has many measures. As a young man, Einstein thought a lot about light and time. One day he had a thought that no one had ever had before. Einstein wondered what a clock would look like if he were riding away from it on a beam of light. He guessed that the clock would appear to stand still. In other words, time would stand still!

Why did Einstein make that guess? Imagine that a clock reads exactly 2 P.M. You can see that because light shining off the clock shows the position of the hands. The light travels to your eye and your brain reads, "2 P.M." A second later, the hand on the clock moves to one second after 2 P.M. Light is still bouncing off the clock. So another beam of light carries a new message to your eye. Now your brain reads, "one second after 2:00 P.M." Beams of light travel so fast that you can read each message instantly. But imagine riding on the beam of light that carries the "2 P.M." message. The other beam of light—the one carrying the message "one second after 2 P.M."—would never catch up with you. So for you, the clock would always read "2 P.M."

Using experiments, Einstein proved his guess was right. He confirmed that speed slows down the passage of time. In that sense, everyone has already done at least a tiny bit of time traveling. You have done it each time you have ridden in a car or plane. Such time travel, however, is far too slight to notice.

Now think like Einstein. Suppose you go very, very fast. In fact, you go almost the speed of light. (Light travels at about 186,000 miles per second! That's more than a million times faster than a jet plane!) At that speed, you'd find that time really does slow down. Messages from other beams of light would catch up with you, but only after a long chase.

Because speed slows down time, you would age slowly as you zip through space. Meanwhile, back on Earth, time would pass as it always does. While you'd be getting five years older, people on Earth might be getting 205 years older. If you return to Earth after your five years, you would find you had traveled two hundred years into the future!

Such time travel is not feasible yet. The fastest spaceships can go only a few thousand miles an hour. We haven't found a way to go any faster. In theory, however, there is a way we could do it. We could use what is called a black hole. A black hole, if it really exists, is a gigantic star that has used up all its fuel. It has collapsed into itself, becoming very small. The gravitational pull from a black hole would be immense. It would be so great, in fact, that everything passing by it—even light beams—would get sucked in. Things would get trapped in a wild funnel that looks a bit like a tornado. The winds in this funnel would be close to the speed of light.

If a spaceship could approach the funnel at just the right angle, the black hole might act as a slingshot. It could whip the spaceship around and send it flying back out through space at a super-high speed. (Of course, the pilot would have to be very careful. He or she could not fly too close to the black hole. Otherwise, the whole spaceship would get pulled in and compressed to less than the size of a grain of sand!)

Now imagine that you want to travel back into the past. That would be even harder to do. You would have to catch up with light beams carrying messages from long ago. To do that, you'd have to travel faster than the speed of light. That is not possible. Nothing can go faster than a light beam. Still, some scientists think there's a way to get around that problem. They suggest taking a shortcut through space. That way, a traveler might be able to catch up with some old beams of light. Scientists have a picture in their minds of what this kind of shortcut would look like. They have even given it a name. They call it a "wormhole." No one knows if wormholes exist. But if they do, travelers might someday use them to jump back in time.

Before you get too excited about traveling to the past, think about some of the questions it would raise. Suppose you traveled back to April 14, 1865. That was the day President Abraham Lincoln was shot. Could you prevent the assassination? Suppose you did. How would that change the course of American history?

Scientists often put the questions in personal terms. Suppose you time-travel back sixty years, to the days when your grandmother is a young woman. Your mother has not yet been born. If you somehow stop your grandmother from meeting your grandfather, where does that leave you? Now your mother won't be born. Does that mean you will cease to exist?

People love to fantasize about time travel. The question is this: Can we really find a way to do it? Will it remain just a dream, carried out only in books or at the movies? Or will we someday be able to fly off into the future and back into the past? Only time will tell.

If you have been timed while reading this selection, enter your reading time on the chart below. Then turn to the Words-per-Minute table on page 111 and look up your reading speed (words per minute). Enter your reading speed on the graph on page 113.

READING TIME: Unit 6

__________ : __________

Minutes *Seconds*

How Well Did You Read?

- *Complete the four exercises that follow. The directions for each exercise will tell you how to mark your answers.*
- *When you have finished all four exercises, use the answer key on page 107 to check your work. For each right answer, put a check mark (✓) on the line beside the box. For each wrong answer, write the correct answer on the line.*
- *Follow the directions after each exercise to find your scores.*

FINDING THE MAIN IDEA

A good main idea statement answers two questions: it tells *who* or *what* is the subject of the story, and it answers the understood question *does what?* or *is what?* Look at the three statements below. One expresses the main idea of the story you just read. Another statement is *too broad*; it is vague and doesn't tell much about the topic of the story. The third statement is *too narrow*; it tells about only one part of the story.

Match the statements with the three answer choices below by writing the letter of each answer in the box in front of the statement it goes with.

M—Main Idea B—Too Broad N—Too Narrow

___ ☐ 1. Some scientists have explored the idea of time travel and believe that it may be possible one day.

___ ☐ 2. Scientists are fascinated by the idea of time travel.

___ ☐ 3. Albert Einstein proved that speed slows down the passage of time.

___ Score 15 points for a correct *M* answer.
___ Score 5 points for each correct *B* or *N* answer.

___ TOTAL SCORE: Finding the Main Idea

B RECALLING FACTS

How well do you remember the facts in the story you just read? Put an *x* in the box in front of the correct answer to each of the multiple-choice questions below.

1. Light travels at the speed of
   - ___ ☐ a. 186,000 feet per second.
   - ___ ☐ b. 186,000 miles per hour.
   - ___ ☐ c. 186,000 miles per second.

2. A black hole is a star that
   - ___ ☐ a. has used up all its fuel.
   - ___ ☐ b. is getting ready to explode.
   - ___ ☐ c. moves faster than the speed of light.

3. Near a black hole, a spaceship could
   - ___ ☐ a. explode.
   - ___ ☐ b. get pulled in and compressed.
   - ___ ☐ c. lose speed and stop moving.

4. To catch up with light beams carrying messages from long ago, you would need to travel
   - ___ ☐ a. the speed of light.
   - ___ ☐ b. the speed of sound.
   - ___ ☐ c. faster than the speed of light.

5. A shortcut through space is sometimes called a
   - ___ ☐ a. funnel.
   - ___ ☐ b. wormhole.
   - ___ ☐ c. slingshot.

Score 5 points for each correct answer.

___ TOTAL SCORE: Recalling Facts

C MAKING INFERENCES

When you use information from the text and your own experience to draw a conclusion that is not directly stated in the text, you are making an *inference.*

Below are five statements that may or may *not* be inferences based on the facts of the story. Write the letter *C* in the box in front of each statement that is a correct inference. Write the letter *F* in front of each faulty inference.

C—Correct Inference F—Faulty Inference

___ ☐ 1. Most scientists ignore the work of Albert Einstein.

___ ☐ 2. Only the finest pilots should fly their ships near black holes.

___ ☐ 3. Someday spaceships will go faster than the speed of light.

___ ☐ 4. If a mother began to travel at almost the speed of light, she could become younger than her daughter who stayed on Earth.

___ ☐ 5. In theory, changing past events would probably have no effect on the present or the future.

Score 5 points for each correct *C* or *F* answer.

___ TOTAL SCORE: Making Inferences

D USING WORDS PRECISELY

Each numbered sentence below contains an underlined word or phrase from the story you have just read. Under the sentence are three definitions. One is a *synonym*, a word that means the same or almost the same thing as the underlined word: *big* and *large* are synonyms. One is an *antonym*, a word that has the opposite or nearly opposite meaning: *love* and *hate* are antonyms. One is an unrelated word; it has a completely *different* meaning than the underlined word. Match the definitions with the three answer choices by writing the letter that stands for each answer in the box in front of the definition it goes with.

S—Synonym A—Antonym D—Different

1. Beams of light travel so fast that you can read each message instantly.

____ ☐ a. slowly
____ ☐ b. immediately
____ ☐ c. clearly

2. Does that mean that you cease to exist?

____ ☐ a. stop
____ ☐ b. begin
____ ☐ c. want

3. He confirmed that speed slows down the passage of time.

____ ☐ a. guessed
____ ☐ b. disproved
____ ☐ c. proved

4. Such time travel is not feasible yet.

____ ☐ a. possible
____ ☐ b. impossible
____ ☐ c. planned

5. The gravitational pull from a black hole would be immense.

____ ☐ a. collapsed
____ ☐ b. huge
____ ☐ c. small

____ Score 3 points for a correct *S* answer.
____ Score 1 point for each correct *A* or *D* answer.

____ TOTAL SCORE: Using Words Precisely

- *Enter the total score for each exercise in the spaces below. Add the scores to find your Critical Reading Score. Then record your Critical Reading Score on the graph on page 114.*

________ Finding the Main Idea
________ Recalling Facts
________ Making Inferences
________ Using Words Precisely

________ CRITICAL READING SCORE: Unit 6

In the movie Jurassic Park, *scientists find a way to use DNA from ancient dinosaurs to produce new dinosaurs. Animation and special effects enable humans to share the screen with frighteningly realistic dinosaurs. Is this plot a far-fetched fantasy of a writer's imagination? Viewers may be surprised to learn that many scientists agree that bringing back dinosaurs is possible. Even if they were managed more wisely than in the movie, however, dinosaurs in today's world would present many practical problems.*

CAN WE BRING BACK THE DINOSAURS?

The last dinosaur died about 65 million years ago. Yet dinosaurs continue to be part of our lives. Toy stores are filled with plastic models of them. Book stores stock many books about them. Cartoons and children's TV shows often feature friendly dinosaurs. Think about your own childhood. What was the first really big word you learned? There's a good chance it was either *Tyrannosaurus* or *Triceratops*.

In 1993, dinosaur fans flocked to a movie called *Jurassic Park*. It was a story about dinosaurs coming back to life in modern times. In the film's first year, people around the world paid $340 million to be scared half out of their minds. There had been many dinosaur movies before. But all those dinosaurs looked phony. *Jurassic Park* was different. The dinosaurs in this movie seemed real. They weren't, of course, but they certainly looked like living creatures.

Jurassic Park seemed real for another reason, as well. The story was based on science. Some researchers think the ideas presented in the movie could actually be carried out. In other words, dinosaurs *could* be brought back to life!

The key to doing it would be to get hold of some dinosaur DNA. DNA is like the blueprint of a creature's genes. It holds the code that makes a cat a cat or a frog a frog. DNA is found in all creatures. It is present in everything from spiders to whales to humans. Dinosaurs, too, had DNA. But how can you get DNA from creatures who died so long ago?

It isn't easy. But here is how some scientists think it might be done. An ancient mosquito bites a dinosaur. The mosquito then gets caught in a sticky tree sap called *amber*. The dripping amber completely covers the unlucky mosquito. In time, this amber dries, turning hard as a rock. That keeps the mosquito's body sealed off from weather and decay.

Now flash forward to the present. Scientists find the amber. They crack it open to reach the mosquito. Using a tiny needle, they extract dinosaur blood from the mosquito's stomach. They manage to recover DNA from the dinosaur's blood cells. With this DNA, scientists re-create living dinosaurs.

That's the theory. But there are still a couple of major problems to be solved.

For one thing, the dinosaur's DNA would not be complete. It would have decayed a bit when it first landed in the mosquito's stomach. Computers would have to be used to guess at what is missing. Then scientists would have to put the dinosaur DNA into a cell from some other creature. If they were lucky, this cell might grow into a baby dinosaur.

So far this has happened only in the movies. But let's suppose it really does happen someday. Then what would we do? Could we keep dinosaurs in a wildlife park like the one shown in *Jurassic Park*? How would we feed them? And how could we be sure they would not attack us?

Some zookeepers say we could train dinosaurs the way we train other large animals. Denver Zoo elephant keeper Liz Hooten says, "You can teach any animal." She says it's done by offering food when an animal does the right thing. Over time the dinosaur would do the right thing in order to get fed. But trainers would have to be very careful. After all, zookeepers are killed every year by the animals they already have. As Hooten warns, "The trick would be not to allow the dinosaurs to associate [humans] too closely with food." If they did, people might end up as a tasty dinosaur snack.

Dinosaurs love to roam. So they would need plenty of space. No zoo today could hold them. Big dinosaurs would need their own wildlife park—a very large wildlife park. In the movie *Jurassic Park*, they were confined to a small island off the coast of Costa Rica. In real life, they would eat up all the food on such an island in a very short time. No, the dinosaurs would need a much larger place. Dinosaur expert James Farlow has suggested the entire country of New Zealand. "Of course, we would have to move the people," he adds.

Getting the right food for dinosaurs would be another problem. Many of the plants that the old dinosaurs ate don't exist anymore, and dinosaurs might not like today's plants. Then there is the problem of disease. Dinosaurs today would face all sorts of germs they have never been exposed to before. So there is no guarantee that we could keep the dinosaurs healthy. They might catch a cold and die. This is a real problem with, say, a forty-foot-long *Tyrannosaurus rex*. How do you get close enough to take its temperature?

Another health problem would be the dinosaurs' toenails. Zookeepers today trim the toenails of elephants. Because these animals don't exercise enough, their toenails grow too long. If dinosaurs are kept in special zoos, someone might have to trim their toenails too. Are there any volunteers to trim the toenails of a *Velociraptor*?

No one knows for sure if dinosaurs will really be brought back to life. But there is no law or principle that says it can't happen. And everything that is not strictly ruled out is possible. So you may someday visit a real Jurassic Park. But if you do, remember this simple rule: "Don't Feed the Animals!" Or, to be more accurate, "Don't Let the Animals Feed on You!"

If you have been timed while reading this selection, enter your reading time on the chart below. Then turn to the Words-per-Minute table on page 111 and look up your reading speed (words per minute). Enter your reading speed on the graph on page 113.

READING TIME: Unit 7

_______________ : _______________

Minutes *Seconds*

How Well Did You Read?

- *Complete the four exercises that follow. The directions for each exercise will tell you how to mark your answers.*
- *When you have finished all four exercises, use the answer key on page 107 to check your work. For each right answer, put a check mark (✓) on the line beside the box. For each wrong answer, write the correct answer on the line.*
- *Follow the directions after each exercise to find your scores.*

A FINDING THE MAIN IDEA

A good main idea statement answers two questions: it tells *who* or *what* is the subject of the story, and it answers the understood question *does what?* or *is what?* Look at the three statements below. One expresses the main idea of the story you just read. Another statement is *too broad*; it is vague and doesn't tell much about the topic of the story. The third statement is *too narrow*; it tells about only one part of the story.

Match the statements with the three answer choices below by writing the letter of each answer in the box in front of the statement it goes with.

M—Main Idea   B—Too Broad   N—Too Narrow

___ ☐ 1. It is possible, though unlikely, that dinosaurs can be brought back to life by using dinosaur DNA.

___ ☐ 2. If dinosaurs came back to the modern world, they would need a very large area in which to live.

___ ☐ 3. The idea of bringing dinosaurs back to life is exciting for everyone.

___ Score 15 points for a correct *M* answer.
___ Score 5 points for each correct *B* or *N* answer.

___ TOTAL SCORE: Finding the Main Idea

B RECALLING FACTS

How well do you remember the facts in the story you just read? Put an *x* in the box in front of the correct answer to each of the multiple-choice questions below.

1. The dinosaurs in *Jurassic Park* were different from those in other movies because they
   ____ ☐ a. sounded frightening.
   ____ ☐ b. looked real.
   ____ ☐ c. were real.

2. The key to bringing back dinosaurs is to get hold of
   ____ ☐ a. dinosaur DNA.
   ____ ☐ b. a mosquito fossil.
   ____ ☐ c. amber from an ancient tree.

3. To make a new dinosaur, scientists would put dinosaur DNA into a
   ____ ☐ a. radioactive test tube.
   ____ ☐ b. mosquito's stomach.
   ____ ☐ c. cell from another creature.

4. According to one zookeeper, to train animals, you
   ____ ☐ a. punish them for misbehaving.
   ____ ☐ b. give them food when they do the right thing.
   ____ ☐ c. treat them with kindness.

5. One expert thinks modern dinosaurs could live in
   ____ ☐ a. Greenland.
   ____ ☐ b. Costa Rica.
   ____ ☐ c. New Zealand.

Score 5 points for each correct answer.

____ TOTAL SCORE: Recalling Facts

C MAKING INFERENCES

When you use information from the text and your own experience to draw a conclusion that is not directly stated in the text, you are making an *inference*.

Below are five statements that may or may *not* be inferences based on the facts of the story. Write the letter *C* in the box in front of each statement that is a correct inference. Write the letter *F* in front of each faulty inference.

C—Correct Inference F—Faulty Inference

____ ☐ 1. The movie *Jurassic Park* was a bigger moneymaker than most films.

____ ☐ 2. The DNA of all creatures is exactly the same.

____ ☐ 3. Mosquitoes and dinosaurs once lived at the same time.

____ ☐ 4. Dinosaurs ate only plants; they never ate animals.

____ ☐ 5. A good place for a modern dinosaur park would be a small valley between two mountains.

Score 5 points for each correct *C* or *F* answer.

____ TOTAL SCORE: Making Inferences

D USING WORDS PRECISELY

Each numbered sentence below contains an underlined word or phrase from the story you have just read. Under the sentence are three definitions. One is a *synonym*, a word that means the same or almost the same thing as the underlined word: *big* and *large* are synonyms. One is an *antonym*, a word that has the opposite or nearly opposite meaning: *love* and *hate* are antonyms. One is an unrelated word; it has a completely *different* meaning than the underlined word. Match the definitions with the three answer choices by writing the letter that stands for each answer in the box in front of the definition it goes with.

S—Synonym A—Antonym D—Different

1. But all those dinosaurs looked <u>phony</u>.

____ ☐ a. real
____ ☐ b. fake
____ ☐ c. ugly

2. It is <u>present</u> in everything from spiders to whales to humans.

____ ☐ a. puzzling
____ ☐ b. existing
____ ☐ c. absent

3. An <u>ancient</u> mosquito bites a dinosaur.

____ ☐ a. hungry
____ ☐ b. new
____ ☐ c. very old

4. Using a tiny needle, they <u>extract</u> dinosaur blood from the mosquito's stomach.

____ ☐ a. take out
____ ☐ b. look for
____ ☐ c. put in

5. The trick would be not to allow the dinosaurs to <u>associate</u> (humans) too closely with food.

____ ☐ a. enjoy
____ ☐ b. separate
____ ☐ c. connect

____ Score 3 points for a correct *S* answer.
____ Score 1 point for each correct *A* or *D* answer.

____ TOTAL SCORE: Using Words Precisely

- *Enter the total score for each exercise in the spaces below. Add the scores to find your Critical Reading Score. Then record your Critical Reading Score on the graph on page 114.*

________	Finding the Main Idea
________	Recalling Facts
________	Making Inferences
________	Using Words Precisely
________	CRITICAL READING SCORE: Unit 7

Maggots are an early stage of flies. They look like little worms and they eat constantly. You may have seen maggots in a garbage pail or on a dead animal along the road. Probably your reaction was "Yecch!" But if you were a doctor treating certain wounds, you might appreciate the maggots more. Recently, doctors have rediscovered what many ancient healers knew. Because maggots have such a huge appetite for decayed flesh, they can clean away dead cells and infection from a wound or sore. Maggots can be good medicine!

THE HEALING POWER OF MAGGOTS

Have you ever opened a garbage pail and found maggots swarming around inside? How did you feel? Were you grossed out? Did these slimy little bugs make you sick to your stomach? These creatures are pretty disgusting. But hold on. Maggots might not really be so creepy after all. In fact, they could save your life!

Maggots are the larvae of flies. Flies start out as eggs. Adult female flies often lay their eggs on food, garbage, rotting plants, or dead animals. These eggs look like tiny grains of rice. In a couple of days, the eggs hatch and out crawl the maggots. These maggots are pale yellow or white. They spend most of their time eating. After a few days, the growing maggots reach the point where they turn into flies.

It's the eating habits of maggots that make them so helpful. Doctors discovered this during World War I. Wounded soldiers were often stranded on the battlefield for hours and even days. In some cases, their open wounds remained clear. In other cases, their wounds became filled with hundreds of maggots. Doctors found that the soldiers with clear wounds often died. The wounds became infected, sending poison surging through the body. It was the infection more than the wounds themselves that killed these men.

But those soldiers with swarming maggots in their wounds had a better record. Many of them survived! Why? The maggots were little eating machines. They constantly searched the wounds for food. Luckily, these maggots had no desire to eat healthy flesh. They craved only decayed flesh and pus. By eating up the rotten parts of a person's body, they helped to prevent infection. In this way, they saved many lives. The maggots helped in other ways, as well. They released a chemical that killed the germs they didn't eat. Also, by crawling over the good flesh, they gave it a healthy massage.

After the war, doctors began to study maggots. They found some interesting things. First, the use of maggots to heal wounds was not really new. It goes back a long way. The Mayan people of Mexico used maggots more than a thousand years ago. In the 1500s, a French doctor named Ambroise Paré noticed maggots in wounds. He felt at the time that the maggots might be doing some good. Later, Baron D. J. Larrey, another French doctor, praised the work of maggots. He noted that most soldiers were terrified when they saw maggots in their wounds. The soldiers calmed down only when they saw the good these creatures did.

By the 1930s, doctors were using maggots as a standard treatment. Drug companies bred maggots and sold them to hospitals. The maggots all came from blow flies. (Maggots from other flies had to be avoided. They would eat healthy flesh!) An average wound needed about five hundred maggots. These maggots would finish their job in two to five days. Then they would turn into flies and fly away.

The use of maggots ended in the 1940s. New wonder drugs took their place. At the time, people were very excited about these drugs. The drugs were neat and clean and easy to use. They were much more appealing to patients. Doctors, too, found them more pleasant to use.

It is easy to understand why

people would prefer drugs to maggots. But was this good science? The new drugs were very expensive. And as it turned out, they didn't always work that well. By the 1990s, doctors were rethinking their position. Said Dr. Jane Petro, "Maggots are more effective and cheaper than a lot of [costly wonder drugs]." Also, maggots are really good at healing tough bone infections. That's because bones have few blood vessels. Modern drugs, which travel through the bloodstream, can't reach these infections.

And so the lowly maggot has been making a mild comeback. Recently a few doctors have started to use them again. Take the case of Dr. Grady Dugas. One of his patients developed deep sores. The sores got infected. They were especially bad on the patient's feet. Dr. Dugas tried using some wonder drugs to clear the sores. None of the drugs seemed to do any good. He then tried surgery to remove the infected tissue. That didn't work either.

Dr. Dugas felt he might have to amputate the patient's feet. Then he remembered his grandmother. She had suffered from sores back in the 1930s. Her doctors had treated the sores with maggots. Dugas recalled that the maggots had healed the sores. So he ordered a supply of blow fly eggs. He placed them in his patient's sores. The eggs hatched, and the maggots went to work. They wiped out the infection, and the sores healed. The maggots had saved the patient's feet!

Still, maggots aren't as popular as they could be. Some doctors remain squeamish about using them. And, remember, maggots turn into flies. Not many people want to walk into a hospital filled with flies. So the healing power of maggots remains a well-kept secret. As Dr. Petro put it, "It just goes back to the disgust factor."

If you have been timed while reading this selection, enter your reading time on the chart below. Then turn to the Words-per-Minute table on page 111 and look up your reading speed (words per minute). Enter your reading speed on the graph on page 113.

READING TIME: Unit 8		
________	:	________
Minutes		*Seconds*

How Well Did You Read?

- *Complete the four exercises that follow. The directions for each exercise will tell you how to mark your answers.*
- *When you have finished all four exercises, use the answer key on page 107 to check your work. For each right answer, put a check mark (✓) on the line beside the box. For each wrong answer, write the correct answer on the line.*
- *Follow the directions after each exercise to find your scores.*

A FINDING THE MAIN IDEA

A good main idea statement answers two questions: it tells *who* or *what* is the subject of the story, and it answers the understood question *does what?* or *is what?* Look at the three statements below. One expresses the main idea of the story you just read. Another statement is *too broad*; it is vague and doesn't tell much about the topic of the story. The third statement is *too narrow*; it tells about only one part of the story.

Match the statements with the three answer choices below by writing the letter of each answer in the box in front of the statement it goes with.

M—Main Idea B—Too Broad N—Too Narrow

___ ☐ 1. Modern medicine is rediscovering that maggots can wipe out the infection in an open wound or sore.

___ ☐ 2. The use of maggots to heal wounds is not new.

___ ☐ 3. Some ancient methods of curing illnesses have been shown to be as effective as modern drugs.

___ Score 15 points for a correct *M* answer.
___ Score 5 points for each correct *B* or *N* answer.

___ TOTAL SCORE: Finding the Main Idea

B RECALLING FACTS

How well do you remember the facts in the story you just read? Put an *x* in the box in front of the correct answer to each of the multiple-choice questions below.

1. A few days after female flies lay eggs,
   - ___ ☐ a. maggots crawl out of the eggs.
   - ___ ☐ b. baby flies hatch and fly away.
   - ___ ☐ c. the eggs turn pale yellow or white.

2. American doctors recognized the benefits of maggots in healing wounds during
   - ___ ☐ a. the Civil War.
   - ___ ☐ b. World War I.
   - ___ ☐ c. the Korean War.

3. Maggots help to cure wounds by
   - ___ ☐ a. eating the rotten parts of a person's body.
   - ___ ☐ b. keeping patient's thoughts on the maggots.
   - ___ ☐ c. eating healthy flesh.

4. The earliest known use of maggots to heal was
   - ___ ☐ a. more than a thousand years ago.
   - ___ ☐ b. in the 1500s.
   - ___ ☐ c. in the 1700s.

5. Reasons for using maggots today do NOT include
   - ___ ☐ a. the higher cost of new drugs.
   - ___ ☐ b. the fact that maggots are more effective than drugs on bone infections.
   - ___ ☐ c. the squeamish attitude of patients and doctors.

Score 5 points for each correct answer.

___ TOTAL SCORE: Recalling Facts

C MAKING INFERENCES

When you use information from the text and your own experience to draw a conclusion that is not directly stated in the text, you are making an *inference.*

Below are five statements that may or may *not* be inferences based on the facts of the story. Write the letter *C* in the box in front of each statement that is a correct inference. Write the letter *F* in front of each faulty inference.

C—Correct Inference **F—Faulty Inference**

___ ☐ 1. Researchers probably took samples of patients' skin to determine how maggots had helped them.

___ ☐ 2. There may be other examples of "folk medicine" that modern medicine does not yet make use of.

___ ☐ 3. The average person could care for his or her own wounds by applying fly eggs to them.

___ ☐ 4. It would be difficult for a doctor to get the approval of both the patient and the hospital for treatment with maggots.

___ ☐ 5. It is likely that we will hear more about treatment with maggots in the near future.

Score 5 points for each correct *C* or *F* answer.

___ TOTAL SCORE: Making Inferences

D USING WORDS PRECISELY

Each numbered sentence below contains an underlined word or phrase from the story you have just read. Under the sentence are three definitions. One is a *synonym*, a word that means the same or almost the same thing as the underlined word: *big* and *large* are synonyms. One is an *antonym*, a word that has the opposite or nearly opposite meaning: *love* and *hate* are antonyms. One is an unrelated word; it has a completely *different* meaning than the underlined word. Match the definitions with the three answer choices by writing the letter that stands for each answer in the box in front of the definition it goes with.

S—Synonym A—Antonym D—Different

1. Wounded soldiers were often stranded on the battlefield for hours and even days.

____ ☐ a. rescued
____ ☐ b. imprisoned
____ ☐ c. left

2. They craved only decayed flesh and pus.

____ ☐ a. desired
____ ☐ b. rejected
____ ☐ c. observed

3. They released a chemical that killed the germs they didn't eat.

____ ☐ a. described
____ ☐ b. sent out
____ ☐ c. held back

4. Dr. Dugas felt that he might have to amputate the patient's feet.

____ ☐ a. give up on
____ ☐ b. re-attach
____ ☐ c. cut off

5. Some doctors remain squeamish about them.

____ ☐ a. easily upset
____ ☐ b. likely to giggle
____ ☐ c. undisturbed

____ Score 3 points for a correct *S* answer.
____ Score 1 point for each correct *A* or *D* answer.

____ TOTAL SCORE: Using Words Precisely

• *Enter the total score for each exercise in the spaces below. Add the scores to find your Critical Reading Score. Then record your Critical Reading Score on the graph on page 114.*

________ Finding the Main Idea
________ Recalling Facts
________ Making Inferences
________ Using Words Precisely

________ CRITICAL READING SCORE: Unit 8

Fortunetellers and mediums claim that they sense things beyond the normal world. They say they see the future, or that they hear from the dead. They use props such as crystal balls and dark rooms. Almost everyone views such characters as entertainers, without real powers. However, it seems that some people do "see" events that happen far away from them. Sometimes they get glimpses of events they don't understand. These people, called psychics*, can't explain how they obtain their information. But what they know can give clues to police stumped by mysterious crimes.*

PSYCHICS WHO SOLVE CRIMES

On December 3, 1967, Dorothy Allison had a frightening dream. In it, she saw the body of a small boy in a river. Allison tried to forget about the dream. But she couldn't. At last she called the police in her town of Nutley, New Jersey. It turned out that five-year-old Michael Kurcsics had indeed drowned in a local river. But he had fallen into the water two hours after Dorothy Allison's dream.

When Allison called the police, they were still looking for the boy's body. They did not want to be bothered by housewives with wild dreams. The tragedy had been reported in the newspapers. So police figured Allison had learned about it there. But Dorothy Allison knew things that had not been printed in the papers. For instance, she said she could see what the boy was wearing.

Police Officer Donald Vicaro was intrigued. He asked Allison what else she could "see." Allison told him that the boy's shoes were on the wrong feet. She said she saw a number "8" and a school with a fence around it. She also saw a gray house and a factory.

A few weeks later, Michael Kurcsics's body was finally found. It had been washed downstream into a nearby pond. When Officer Vicaro arrived at the scene, he could hardly believe his eyes. There, next to the pond, was Public School Number 8. Around the school was a fence. A factory stood in the distance and so did a gray house. Vicaro checked the boy's body. As Dorothy Allison had predicted, the shoes were on the wrong feet!

Dorothy Allison is a psychic. She seems to have special powers to "see" things. Some of her visions come from far-away places. Some come from the future, others come from the past. Allison is not sure how or why she has these visions. She only knows that she's been getting them since childhood.

After the Kurcsics case, police asked Allison to help solve other crimes. By 1987, she had helped crack hundreds of cases. She does not take any pay for her work. She is just happy when her visions can be put to good use. Often Allison herself does not know what her visions mean. As one detective says, "She may see things backwards, forwards, in the middle. . . . It's up to the police to put the information in some kind of order." Adds Sheriff Dale Dye, "Working with a psychic is like doing a crossword puzzle!"

Allison is not the only psychic who has helped solve crimes. There are hundreds like her. In Delavan, Illinois, police sometimes turn to Greta Alexander. In 1977, Alexander told police where to find two drowning victims. Six years later, she helped out again. A woman had been missing for a month. Alexander told police to go to a wooded spot. It was near the town of Peoria. There, she said, the woman's body would be found near a bridge. A pile of rocks or salt would also be found there. She warned police that the woman's head would be detached from her body. Alexander was right on all counts.

Pennsylvania psychic Nancy Czetli has worked on more than a hundred cases. She has helped solve murders. She has helped find kidnappers and track down burglars. In January of 1988, she was asked to find a seventy-eight-year-old man. He had gone out for a walk but never returned. Police had searched for him for a week. They had used dogs, a helicopter—everything. They had found no trace of him. Yet when Czetli looked at an old photograph of the man, she could sense right away what had happened. He had died from the cold. Czetli pointed out the path he had followed. She led police right to the spot where his body was found.

Texas psychic John Catchings began working with police in 1980. He was asked to help find an eighteen-year-old boy who had disappeared. Catchings felt at once that the boy had been murdered. He asked to hold something that had belonged to the boy. He was given the boy's high school ring. "When I held the ring," Catchings says, "I saw a white house with peeling paint, a trail behind the house, weeds, an old tire, a shoe, a creek."

Police recognized the place Catchings described. It was near the missing boy's home. Catchings announced that the boy's body would be found there. "An old shoe will be the marker," he said. "You'll find the boy's left heel and ankle exposed."

Police had already searched the property once. But they decided to look again. Sure enough, this time they noticed a tree with tires piled up around it. On top of one tire was a sandal. Could that be the shoe Catchings mentioned? When police moved the tires, they saw a heel sticking out of the ground. Digging up the dirt, they found the body of the missing boy.

Psychics work in a variety of ways. Some "see" crimes in their dreams. Some say they can read the mind of a victim. Even after a person has died, they say, his or her thought patterns linger on in the brain. By reading those patterns, they can find out what happened. "I don't become the victim," says Nancy Czetli, "but it's as if I'm standing alongside him."

Many psychics also use psychometry. This involves getting information from objects. Psychics may ask to hold something that belonged to the victim. It could be anything. It could be an old hat or—as in John Catching's famous case—a high school ring. Psychics say they can get vibrations from these things. By feeling the vibrations, they can tell what happened to the object's owner.

Psychics have not convinced everyone. Some people still scoff at them. Dr. Martin Reiser has done studies for the Los Angeles Police Department. He says psychics are of no use to police. But more and more people are echoing the words of Detective Ron Phillips. "I felt it was baloney at first," he said. But then Phillips worked with John Catchings. "He made me believe [in psychics]," Phillips said. "They've got a power there—it gives you goose bumps, really."

If you have been timed while reading this selection, enter your reading time on the chart below. Then turn to the Words-per-Minute table on page 111 and look up your reading speed (words per minute). Enter your reading speed on the graph on page 113.

READING TIME: Unit 9	
______ :	______
Minutes	*Seconds*

How Well Did You Read?

- *Complete the four exercises that follow. The directions for each exercise will tell you how to mark your answers.*
- *When you have finished all four exercises, use the answer key on page 107 to check your work. For each right answer, put a check mark (✓) on the line beside the box. For each wrong answer, write the correct answer on the line.*
- *Follow the directions after each exercise to find your scores.*

A FINDING THE MAIN IDEA

A good main idea statement answers two questions: it tells *who* or *what* is the subject of the story, and it answers the understood question *does what?* or *is what?* Look at the three statements below. One expresses the main idea of the story you just read. Another statement is *too broad;* it is vague and doesn't tell much about the topic of the story. The third statement is *too narrow;* it tells about only one part of the story.

Match the statements with the three answer choices below by writing the letter of each answer in the box in front of the statement it goes with.

M—Main Idea B—Too Broad N—Too Narrow

___ ☐ 1. Psychics have a special knowledge of things that have happened or are about to happen.

___ ☐ 2. Psychics' ability to "see" things that others can't makes them valuable in solving crimes.

___ ☐ 3. Psychic Nancy Czetli has been helpful in solving cases of murder, kidnapping, and burglary.

___ Score 15 points for a correct *M* answer.
___ Score 5 points for each correct *B* or *N* answer.

___ TOTAL SCORE: Finding the Main Idea

B RECALLING FACTS

How well do you remember the facts in the story you just read? Put an *x* in the box in front of the correct answer to each of the multiple-choice questions below.

1. At first, police ignored Dorothy Allison's offer to help because they
   - ____ ☐ a. thought she read about it in the newspaper.
   - ____ ☐ b. had all the clues they needed.
   - ____ ☐ c. didn't like housewives.

2. One amazing detail Allison had "seen" was that
   - ____ ☐ a. the boy was carrying $100.
   - ____ ☐ b. the boy's hair was red.
   - ____ ☐ c. the boy's shoes were on the wrong feet.

3. More police are asking for help from psychics because
   - ____ ☐ a. psychics usually work for free.
   - ____ ☐ b. they want one more way to fight crime.
   - ____ ☐ c. regular ways of solving crimes never work.

4. John Catchings was handed a missing boy's
   - ____ ☐ a. high school ring.
   - ____ ☐ b. team jacket.
   - ____ ☐ c. wallet.

5. Some psychics say that when they hold an object that belongs to a victim, they
   - ____ ☐ a. can hear the voice of the victim.
   - ____ ☐ b. become the victim themselves.
   - ____ ☐ c. get vibrations from the object.

Score 5 points for each correct answer.

____ TOTAL SCORE: Recalling Facts

C MAKING INFERENCES

When you use information from the text and your own experience to draw a conclusion that is not directly stated in the text, you are making an *inference*.

Below are five statements that may or may not be inferences based on the facts of the story. Write the letter *C* in the box in front of each statement that is a correct inference. Write the letter *F* in front of each faulty inference.

C—Correct Inference F—Faulty Inference

____ ☐ 1. Psychic Dorothy Allison is usually thought to be a greedy person.

____ ☐ 2. Psychics may be born with their unusual powers.

____ ☐ 3. Psychics do their best work in their hometowns.

____ ☐ 4. Police departments usually bring psychics in on their cases after usual methods have failed.

____ ☐ 5. It takes clever police work to use the information that psychics give.

Score 5 points for each correct *C* or *F* answer.

____ TOTAL SCORE: Making Inferences

D USING WORDS PRECISELY

Each numbered sentence below contains an underlined word or phrase from the story you have just read. Under the sentence are three definitions. One is a *synonym*, a word that means the same or almost the same thing as the underlined word: *big* and *large* are synonyms. One is an *antonym*, a word that has the opposite or nearly opposite meaning: *love* and *hate* are antonyms. One is an unrelated word; it has a completely *different* meaning than the underlined word. Match the definitions with the three answer choices by writing the letter that stands for each answer in the box in front of the definition it goes with.

S—Synonym A—Antonym D—Different

1. The tragedy had been reported in the newspapers.

___ ☐ a. story
___ ☐ b. comedy
___ ☐ c. sad event

2. Police Officer Donald Vicaro was intrigued.

___ ☐ a. greatly interested
___ ☐ b. stubborn
___ ☐ c. uncaring

3. She warned police that the woman's head would be detached from her body.

___ ☐ a. attached
___ ☐ b. separated
___ ☐ c. different

4. You'll find the boy's left heel and ankle exposed.

___ ☐ a. hidden
___ ☐ b. dirty
___ ☐ c. uncovered

5. Even after a person has died, they say, his or her thought patterns linger on in the brain.

___ ☐ a. read
___ ☐ b. stay
___ ☐ c. leave

___ Score 3 points for a correct *S* answer.
___ Score 1 point for each correct *A* or *D* answer.

___ TOTAL SCORE: Using Words Precisely

• *Enter the total score for each exercise in the spaces below. Add the scores to find your Critical Reading Score. Then record your Critical Reading Score on the graph on page 114.*

______	Finding the Main Idea
______	Recalling Facts
______	Making Inferences
______	Using Words Precisely
______	CRITICAL READING SCORE: Unit 9

Think of a time when you had a flower or leaf that you wanted to keep for a long time. Before you put it away, you treated it specially so it wouldn't decay or rot. The people of ancient Egypt faced the same problem with other valuable items that had once been alive—human bodies. They believed that spirits of dead people still needed their bodies. So, after people died, their bodies were given special treatment to keep them from decaying in their graves. Bodies given this treatment are called **mummies.** *The process of making a mummy involved many steps, including wrapping the body.*

MUMMIES

Have you ever thought of being a mummy for Halloween? You could probably do it just by wrapping yourself in a bunch of white bandages. But making a real mummy is not so easy. First, you need a dead body. And second, you need someone who understands the ancient art of mummy-making.

A mummy is not the same as a skeleton. A skeleton is just bones. But a mummy has bones and skin. Often it has hair, fingernails, and muscles as well. Usually, these softer body parts decay quickly. But that does not happen with mummies. If a mummy has been properly prepared, it can last a long, long time.

Ancient Egyptians were master mummy makers. Some of their mummies are now over three thousand years old. These bodies still have lips, noses, and ears. They have eyelids and toenails. One has red hair. Another has a face twisted in a scream of death made thousands of years ago.

How did the Egyptians make such great mummies? Their secret was to dry each body thoroughly. If a body is totally dry, the flesh won't rot away. Drying a body means getting rid of all the fluids in it. To do this, Egyptians slit open the side of each body. They scooped out much of the insides. They took out the stomach. They took out the liver, intestines, and lungs. They did not take out the heart, however. They believed the heart was the center of wisdom and truth. So they left that alone.

The brain, on the other hand, didn't seem important. Mummy makers stuck a long metal hook up the nose of each body. They cracked through the skull. Then they pulled out the brain and threw it away.

Once Egyptians had cleaned out a body, they washed it with wine. They packed it with a special salt called *natron*. Then they sat back and waited for forty days. During that time, the natron soaked up liquid from the body. By the end of the forty days, no liquid was left.

The drying-out process left bodies as shriveled as prunes. But Egyptians solved that problem too. They stuffed the bodies with cloth or sand. That puffed the skin back up again. Sometimes they put peppercorns up the nose. That helped push the nose back to its original shape. Egyptians also rubbed spices and herbs on the body to mask the smell of death.

Next, mummy makers coated the dried body with a glue called *resin*. As the resin dried, it became hard. It formed a tough coating that protected the body. It made the body waterproof.

Finally, Egyptians wrapped each body in twenty layers of cloth. This was a tricky task. It took many days and required about 150 yards of cloth. Sometimes an ear or toe fell off during the wrapping. But by the time the last layer was put on, the mummy was close to normal size again.

From 2600 B.C. to A.D. 300, Egyptians made millions of mummies. Almost everyone in Egypt who could afford it became one. Egyptians also mummified animals. They made cat mummies and dog mummies. They turned fish, snakes, and birds into mummies. They even made mummies out of grasshoppers and beetles!

There was a reason for this "mummy-mania." Egyptians thought living things needed their bodies after death. They believed dead people went on to the land of the gods. Even dead animals moved on to an "afterlife." Spirits of the dead could make contact with the gods. But those spirits still needed a place to rest at night. They needed to return to their bodies. If their bodies had rotted away, the spirits would have no place to rest. Then the spirits, too, would die.

Ancient Egyptians were not the only ones who believed in life after death. And they were not the only ones who made mummies. Halfway across the world, in the mountains of South America, people did the same thing. By 3000 B.C., people in Peru and Chile had figured out how to preserve dead bodies. They did not dry them with salt. Instead, they set them out in the hot sun. Sometimes they also put them over a fire. The heat and smoke helped to get rid of all liquids. Once the bodies were dry, they were wrapped up and put in baskets. In Egypt, mummies were stretched out flat. But most South American mummies had their knees folded up to their chins.

Mummies are not always thousands of years old. Four hundred years ago, some people in Italy started making mummies. They felt it would help them keep in touch with the spirits of those who had died. These Italians put the bodies in a special room. They left them there for a year. During that time, all fluids drained out of the bodies. Then the bodies were laid in the sun. When they were fully dried, they were dressed in fancy clothes. They were put in underground rooms called *catacombs*.

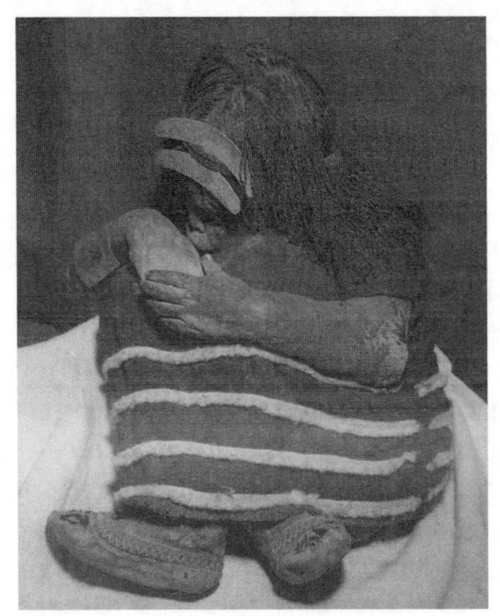

This mummy, found in Chile in 1954, is that of a fifteen-year-old Inca girl. She died about 500 years ago. After her body dried out, her knees were folded up to her chin.

People often visited the mummies in the catacombs. They brought picnic lunches to eat. They talked to the mummies, asking them for advice. Some people even held hands with the mummies as they said prayers.

The last Italian mummy was created in 1920. It was made from the body of a little girl named Rosali Lombardo. Rosali died at the age of two. Her father was a doctor. He knew how to mummify dead bodies. In fact, he had developed a new system for it. Dr. Lombardo used his system on little Rosali. The results were amazing. Rosali's body is perfectly preserved. It has not shriveled at all. In fact, it looks as though Rosali is just taking a nap. No one knows what Dr. Lombardo's system was. He died before sharing it with the world.

Some people think mummies are a good way to honor the dead. But others don't even like to look at them. You might want to keep that in mind if you ever do dress up like a mummy for Halloween.

If you have been timed while reading this selection, enter your reading time on the chart below. Then turn to the Words-per-Minute table on page 111 and look up your reading speed (words per minute). Enter your reading speed on the graph on page 113.

READING TIME: Unit 10		
________	:	________
Minutes		*Seconds*

How Well Did You Read?

- *Complete the four exercises that follow. The directions for each exercise will tell you how to mark your answers.*

- *When you have finished all four exercises, use the answer key on page 107 to check your work. For each right answer, put a check mark (✓) on the line beside the box. For each wrong answer, write the correct answer on the line.*

- *Follow the directions after each exercise to find your scores.*

A FINDING THE MAIN IDEA

A good main idea statement answers two questions: it tells *who* or *what* is the subject of the story, and it answers the understood question *does what?* or *is what?* Look at the three statements below. One expresses the main idea of the story you just read. Another statement is *too broad;* it is vague and doesn't tell much about the topic of the story. The third statement is *too narrow;* it tells about only one part of the story.

Match the statements with the three answer choices below by writing the letter of each answer in the box in front of the statement it goes with.

M—Main Idea B—Too Broad N—Too Narrow

___ ☐ 1. Many people around the world and in different times shared a belief in an afterlife, and this belief affected burial customs.

___ ☐ 2. Although other people have made mummies, too, the process used by ancient Egyptians is the best known.

___ ☐ 3. Many mummies have not only bones and skin, but also hair, fingernails, and muscles.

___ Score 15 points for a correct *M* answer.
___ Score 5 points for each correct *B* or *N* answer.

___ TOTAL SCORE: Finding the Main Idea

B RECALLING FACTS

How well do you remember the facts in the story you just read? Put an *x* in the box in front of the correct answer to each of the multiple-choice questions below.

1. The first step in making a mummy is to
   - ____ □ a. dry the body thoroughly.
   - ____ □ b. remove most inner organs and the brain.
   - ____ □ c. wrap the body in layers of cloth.

2. The Egyptians used a glue called *resin* to
   - ____ □ a. give the dried-out body a natural shape.
   - ____ □ b. glue onto the body any small parts, such as ears and toes, that had fallen off.
   - ____ □ c. waterproof the body.

3. The Egyptians thought that, at death, a spirit
   - ____ □ a. died along with its body.
   - ____ □ b. left the body forever to go to the gods.
   - ____ □ c. visited the gods but returned to its body to rest.

4. Mummies made in South America usually
   - ____ □ a. were dressed in fancy clothes.
   - ____ □ b. had their knees folded up to their chins.
   - ____ □ c. were buried in large wooden boxes.

5. Italians who made mummies during the last four hundred years did so in order to
   - ____ □ a. keep in touch with the spirits of the dead.
   - ____ □ b. display the mummies in museums.
   - ____ □ c. save the bodies of the dead for future cures.

Score 5 points for each correct answer.

____ TOTAL SCORE: Recalling Facts

C MAKING INFERENCES

When you use information from the text and your own experience to draw a conclusion that is not directly stated in the text, you are making an *inference*.

Below are five statements that may or may *not* be inferences based on the facts of the story. Write the letter *C* in the box in front of each statement that is a correct inference. Write the letter *F* in front of each faulty inference.

C—Correct Inference F—Faulty Inference

____ □ 1. In ancient Egypt, the process of making a human body into a mummy cost a great deal.

____ □ 2. Making mummies is easier in desert areas than in rainy lands.

____ □ 3. The people of South America got the idea of making mummies from the people of Egypt.

____ □ 4. Ancient Egyptians would think that a Halloween mummy costume shows disrespect for the dead.

____ □ 5. Doctors in ancient Egypt had a good idea of how each part of the body worked and what it did for the living person.

Score 5 points for each correct *C* or *F* answer.

____ TOTAL SCORE: Making Inferences

D USING WORDS PRECISELY

Each numbered sentence below contains an underlined word or phrase from the story you have just read. Under the sentence are three definitions. One is a *synonym*, a word that means the same or almost the same thing as the underlined word: *big* and *large* are synonyms. One is an *antonym*, a word that has the opposite or nearly opposite meaning: *love* and *hate* are antonyms. One is an unrelated word; it has a completely *different* meaning than the underlined word. Match the definitions with the three answer choices by writing the letter that stands for each answer in the box in front of the definition it goes with.

S—Synonym A—Antonym D—Different

1. Usually, these softer body parts <u>decay</u> quickly.

____ ☐ a. grow
____ ☐ b. shake
____ ☐ c. rot

2. Their secret was to dry each body <u>thoroughly</u>.

____ ☐ a. completely
____ ☐ b. at a high altitude
____ ☐ c. imperfectly

3. The drying-out process left bodies as <u>shriveled</u> as prunes.

____ ☐ a. colorful
____ ☐ b. shrunken and wrinkled
____ ☐ c. swollen

4. Even dead animals moved on to an "<u>afterlife</u>."

____ ☐ a. nothingness
____ ☐ b. life after death
____ ☐ c. judgment

5. By 3000 B.C., people in Peru and Chile had figured out how to <u>preserve</u> dead bodies.

____ ☐ a. keep
____ ☐ b. display
____ ☐ c. destroy

____ Score 3 points for a correct *S* answer.
____ Score 1 point for each correct *A* or *D* answer.

____ TOTAL SCORE: Using Words Precisely

• *Enter the total score for each exercise in the spaces below. Add the scores to find your Critical Reading Score. Then record your Critical Reading Score on the graph on page 114.*

________ Finding the Main Idea
________ Recalling Facts
________ Making Inferences
________ Using Words Precisely

________ CRITICAL READING SCORE: Unit 10

Group Three

Imagine that you have been seriously injured in a car crash. Sirens grow louder as ambulances race toward the scene. But you stop breathing. You are conscious of rising above your body. Free of pain, you begin floating through a tunnel toward a distant bright light. You feel peaceful and content. You glance back at the crash scene and watch a medic begin CPR on your body. The bright light fades, and you hear someone yell excitedly, "This victim is breathing again!" Alive once more, you join the millions of other Americans who have had a Near-Death Experience.

NEAR-DEATH EXPERIENCES

At 1:38 on a Tuesday afternoon, Richard Selzer died. He had lain in a coma for 23 days, and, finally, his heart stopped beating. The doctor and nurses did all they could to get it going again. They gave Selzer electric shocks. They injected medicine right into his chest. But at 1:38 P.M. on April 23, 1991, the doctor declared, "This man is dead." Imagine everyone's surprise when, ten minutes later, Richard Selzer began breathing again.

No one knows just how Selzer did it. During the minutes he was "dead," he had no heartbeat. He drew no breath. Yet he can describe what went on in the hospital room during those minutes. In 1993, he laid it all out in a book titled, *Raising the Dead.*

In the book, Selzer details the feeling of being outside his body. Looking down at the hospital bed from above, he could see what was happening. He saw the movements made by nurses after he was declared dead. Just before coming back to life, he heard the beating of wings. Then he felt a veil lifting from his face. That was when he began to breathe again.

Richard Selzer is not the only one who has had this kind of brush with death. In a 1982 poll, 8 million Americans reported that they had had Near-Death Experiences, or NDEs. JoDee Chenaur had one when her heart stopped during surgery. Beverly Brodsky had one after a bad motorcycle crash. Young Stuart Twemlow had one when he fell into a washtub and almost drowned.

What happens to people during the time they are "dead"? Different people report different things. Still, patterns have formed. Like Richard Selzer, many say they leave their bodies. Grace Bubulka put it this way: "I was out of my body and out of pain. I was up on the ceiling in a corner of the room, looking down, watching doctors and nurses rush around frantically as they worked to save my life." Laurelynn Martin said, "I remember just floating up through darkness." Writes David Wheeler, "I felt myself moving away from my physical body. . . . I started to float just a little distance above my body."

Next, people often say, they move through a tunnel. They travel toward a bright light. Along the way, they may meet dead relatives. Or they may look back over their whole life. For most people, these moments are pleasant. According to David Wheeler, "I was not frightened. It was a good feeling." Others agree. One woman said, "What I saw while I was dead was so beautiful." Another man claimed his NDE "was the most relaxing and joyful experience of my life."

For a few, though, NDEs bring terror. These people feel they are entering a giant void. They see nothing but dark, empty space. Janine Eharrat felt she was "falling into a deep well. The fall never seemed to end. I was alone in a strange and unfamiliar world. . . ."

People who have had Near-Death Experiences—good or bad—feel their journeys were somehow interrupted. Some say they were pulled back to their bodies. Others say they were sent back against their will. In either case, their hearts resumed beating and they were once again "alive."

Are people making up these stories? Perhaps. But it's hard to believe that 8 million folks are lying. So what is going on? Do people really leave their bodies during NDEs? Some scientists say no. They say people may think they are floating. But that feeling is caused by lack of oxygen to the brain. Dr. Bruce Greyson disagrees. Greyson is a college professor. He has spent twenty years studying NDEs. He points out that lack of oxygen causes confusion and panic. NDEs, on the other hand, bring calm, clear thoughts.

Could NDEs simply be dreams? Greyson throws out that theory, too. Dreams, he says, don't change people's lives. Near-Death Experiences do. NDEs leave people happier and less fearful. After NDEs, people tend to focus on helping others. They may give up high-paying jobs to do work that pays less but is more satisfying.

Dr. Sherwin Nuland is not convinced. He thinks NDEs are caused by chemicals in the brain. He says these chemicals are sent out in times of shock. But Nuland can't explain all parts of NDEs. Sometimes, for instance, people pick up information while they float outside their bodies. Greyson tells of a woman who "died" for a short time. While doctors worked to bring her back, Greyson talked with the woman's roommate. Later, the woman could describe that whole conversation. Said Greyson, "Even if she had been conscious, she couldn't possibly have overheard. We were too far away."

Sooner or later, we will all find out for ourselves what happens when we die. But people who have had NDEs urge us not to rush things. They enjoyed their contact with death. But they still want to continue their lives here on Earth. As Dr. Greyson notes, an NDE does not make someone suicidal. "On the contrary," he says, "it makes life more attractive."

If you have been timed while reading this selection, enter your reading time on the chart below. Then turn to the Words-per-Minute table on page 112 and look up your reading speed (words per minute). Enter your reading speed on the graph on page 113.

READING TIME: Unit 11

__________ : __________

Minutes *Seconds*

How Well Did You Read?

- *Complete the four exercises that follow. The directions for each exercise will tell you how to mark your answers.*
- *When you have finished all four exercises, use the answer key on page 108 to check your work. For each right answer, put a check mark (✓) on the line beside the box. For each wrong answer, write the correct answer on the line.*
- *Follow the directions after each exercise to find your scores.*

A FINDING THE MAIN IDEA

A good main idea statement answers two questions: it tells *who* or *what* is the subject of the story, and it answers the understood question *does what?* or *is what?* Look at the three statements below. One expresses the main idea of the story you just read. Another statement is *too broad*; it is vague and doesn't tell much about the topic of the story. The third statement is *too narrow*; it tells about only one part of the story.

Match the statements with the three answer choices below by writing the letter of each answer in the box in front of the statement it goes with.

M—Main Idea B—Too Broad N—Too Narrow

____ ☐ 1. Ten minutes after being declared dead, Richard Selzer began to breathe again and later told of his Near-Death Experience.

____ ☐ 2. What people report about their Near-Death Experiences is fascinating.

____ ☐ 3. Sometimes people who seemed to have died come back to life and tell about their Near-Death Experiences.

____ Score 15 points for a correct *M* answer.
____ Score 5 points for each correct *B* or *N* answer.

____ TOTAL SCORE: Finding the Main Idea

B RECALLING FACTS

How well do you remember the facts in the story you just read? Put an *x* in the box in front of the correct answer to each of the multiple-choice questions below.

1. One experience that many people report during the time they are "dead" is
   - ____ ☐ a. floating out of their bodies.
   - ____ ☐ b. falling asleep.
   - ____ ☐ c. hearing the sound of the ocean.

2. When Richard Selzer was "dead," he saw
   - ____ ☐ a. heaven.
   - ____ ☐ b. what was happening near his hospital bed.
   - ____ ☐ c. his home and family.

3. In their NDEs, many people report that they meet
   - ____ ☐ a. kind strangers.
   - ____ ☐ b. angels.
   - ____ ☐ c. dead relatives.

4. Some scientists say that the floating feeling that often comes with an NDE is caused by
   - ____ ☐ a. panic and fear.
   - ____ ☐ b. a lack of oxygen to the brain.
   - ____ ☐ c. electrical discharges.

5. Dr. Bruce Greyson believes that NDEs are not simply dreams because
   - ____ ☐ a. dreams don't change people's lives.
   - ____ ☐ b. people forget dreams.
   - ____ ☐ c. many people dream in color.

Score 5 points for each correct answer.

____ TOTAL SCORE: Recalling Facts

C MAKING INFERENCES

When you use information from the text and your own experience to draw a conclusion that is not directly stated in the text, you are making an *inference*.

Below are five statements that may or may *not* be inferences based on the facts of the story. Write the letter *C* in the box in front of each statement that is a correct inference. Write the letter *F* in front of each faulty inference.

C—Correct Inference F—Faulty Inference

____ ☐ 1. You will probably have a Near-Death Experience at least once in your life.

____ ☐ 2. If many people say the same thing, it must be true.

____ ☐ 3. Reading about others' Near-Death Experiences can make people less afraid of their own deaths.

____ ☐ 4. The only cause of confusion and panic is lack of oxygen to the brain

____ ☐ 5. Someone who has experienced an NDE is likely to be sad and depressed.

Score 5 points for each correct *C* or *F* answer.

____ TOTAL SCORE: Making Inferences

D USING WORDS PRECISELY

Each numbered sentence below contains an underlined word or phrase from the story you have just read. Under the sentence are three definitions. One is a *synonym*, a word that means the same or almost the same thing as the underlined word: *big* and *large* are synonyms. One is an *antonym*, a word that has the opposite or nearly opposite meaning: *love* and *hate* are antonyms. One is an unrelated word; it has a completely *different* meaning than the underlined word. Match the definitions with the three answer choices by writing the letter that stands for each answer in the box in front of the definition it goes with.

S—Synonym **A—Antonym** **D—Different**

1. They injected medicine right into his chest.

___ ☐ a. breathed
___ ☐ b. inserted
___ ☐ c. removed

2. I was . . . watching doctors and nurses rush around frantically as they worked to save my life.

___ ☐ a. sadly
___ ☐ b. calmly
___ ☐ c. excitedly

3. For a few, though, NDEs bring terror.

___ ☐ a. intense fear
___ ☐ b. curiosity
___ ☐ c. relaxation

4. People who have had Near-Death Experiences—good or bad—feel their journeys were somehow interrupted.

___ ☐ a. continued
___ ☐ b. stopped
___ ☐ c. misunderstood

5. Greyson throws out that theory, too.

___ ☐ a. proven fact
___ ☐ b. study
___ ☐ c. guess

___ Score 3 points for a correct *S* answer
___ Score 1 point for each correct *A* or *D* answer

___ TOTAL SCORE: Using Words Precisely

- *Enter the total score for each exercise in the spaces below. Add the scores to find your Critical Reading Score. Then record your Critical Reading Score on the graph on page 114.*

______	Finding the Main Idea
______	Recalling Facts
______	Making Inferences
______	Using Words Precisely
______	CRITICAL READING SCORE: Unit 11

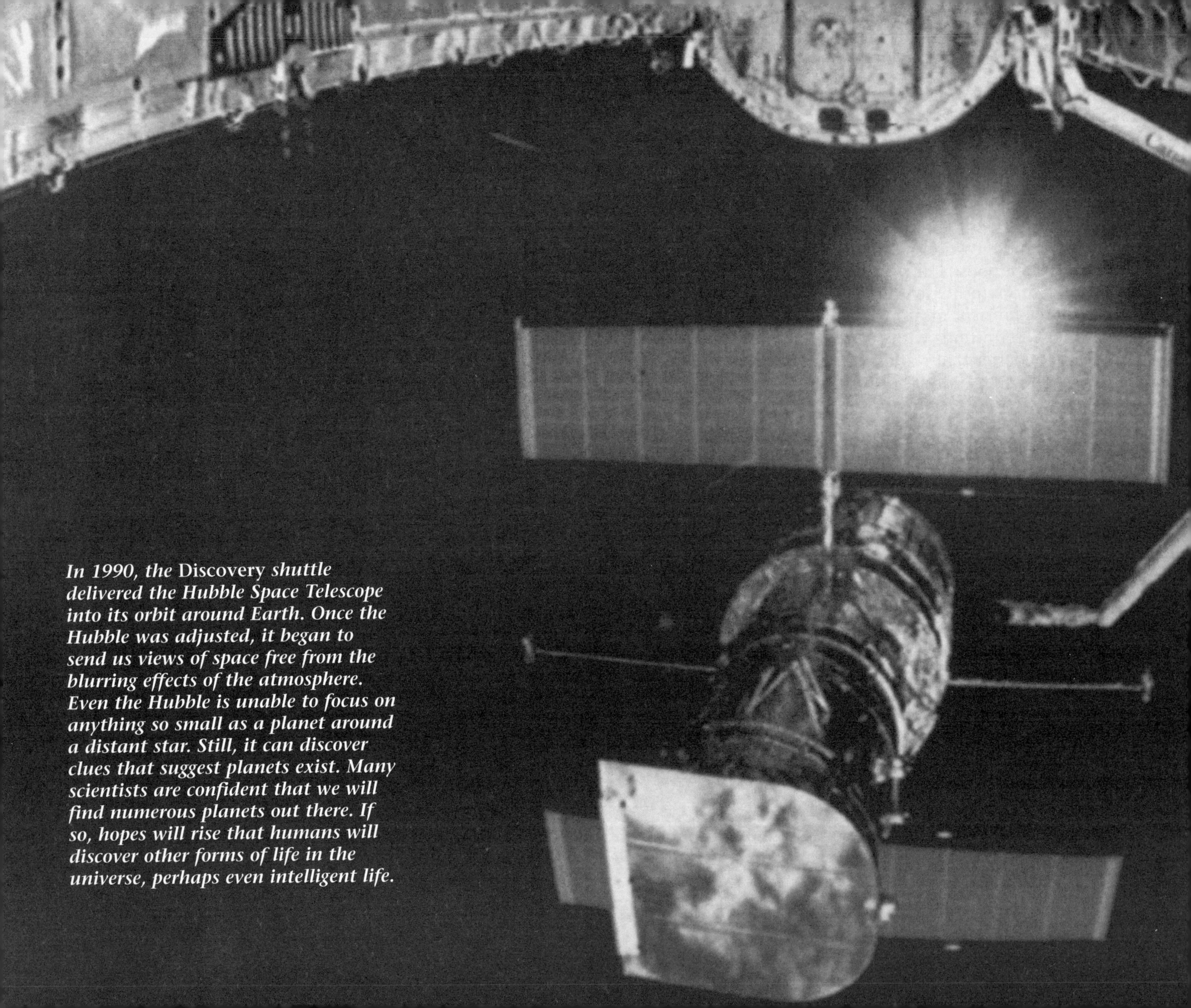

In 1990, the Discovery *shuttle delivered the Hubble Space Telescope into its orbit around Earth. Once the Hubble was adjusted, it began to send us views of space free from the blurring effects of the atmosphere. Even the Hubble is unable to focus on anything so small as a planet around a distant star. Still, it can discover clues that suggest planets exist. Many scientists are confident that we will find numerous planets out there. If so, hopes will rise that humans will discover other forms of life in the universe, perhaps even intelligent life.*

IS ANYONE OUT THERE?

Is there intelligent life elsewhere in the universe? Some people would answer that question with a loud, "Yes!" In fact, some people would say that space travelers from other worlds visit us all the time. As proof, these people would point to UFOs (Unidentified Flying Objects). There are thousands of UFO sightings every year. Believers say UFOs are really spaceships from other planets. So for UFO fans, there is no question. They are sure someone else is "out there" and, in fact, is watching us all the time.

Yet despite all the sightings, most scientists do not believe in UFOs. There are a couple of reasons for this. For one thing, how would creatures from some other planet know about us? We have been sending radio signals to outer space for only a few years. These signals have not had time to reach any distant planets. Our nearest neighbors may be hundreds of light-years away. (A light-year measures how far light travels in one year. The speed of light is 186,000 miles per second.) So if we do have neighbors, they won't get our radio signals for years to come.

Some people think aliens could have noticed us even without radio signals. If that's the case, they say, couldn't these aliens drop by for a visit from time to time? It's not likely. Certainly humans have not figured out a way to whiz from one solar system to another. Again, it's a distance problem. The nearest star is more than four light-years away. It would take our best spaceship one hundred thousand years to get there! And most stars are much, much farther away.

So if aliens were going to visit us, they'd have to be a lot smarter than we are. They would have to figure out how to fly close to the speed of light. Some distant civilization might have the skills to do that. But, again, most scientists doubt it. Besides, suppose that some life form is that advanced. Why would these creatures want to visit Earth? Compared to them, humans would seem pretty simple-minded. So if aliens were going to zip off to some other planet, chances are it wouldn't be ours.

Given all this, it seems safe to conclude that UFOs are not for real.

Little green people in flying saucers are probably not flashing through the sky every year to check us out. Does that mean that there is no intelligent life anywhere else in the universe? Not at all! Most scientists believe there is lots of intelligent life out there. Just look at the facts. In our galaxy alone—the Milky Way—there are about 400 billion stars. There is a strong chance that many of these stars have planets that can support life. Scientists have made a rough guess about how many. They figure that there may be as many as ten thousand civilizations in the Milky Way. Now consider that there are at least 400 billion galaxies! Surely these other galaxies also contain planets that could support life.

It's frustrating to think that we might never see the life forms that inhabit other planets. But couldn't we at least talk to them? Scientists say that is possible. It could be done using radio waves. These waves travel at the speed of light. Even here, though, there are some problems. Imagine radio signals coming from a planet that is ten thousand light-years away. By the time we get these signals, the creatures who sent them might no longer exist. After all, it would have taken the signals ten thousand years to reach Earth. A lot can happen in that time. Think about our own civilization. Will the human race be here ten thousand years from now? Or will some disease have killed us all? Will we have wiped ourselves out with wars or pollution? No one knows the answer. In fact, there are no guarantees we'll last even another hundred years. So by the time creatures on another planet receive our radio signals, we might be long gone.

There's one more thing to keep in mind. The life that exists on another planet might not look anything at all like human life. Scientists urge us to stop thinking in terms of "little green men." In truth, we have no idea what other forms of intelligent life would look like. Like us, they would most likely be made up of atoms and molecules. But beyond that, it's anyone's guess. They could be as different from us as we are from alligators. Also, the idea of flying saucers is ours, not theirs. If aliens did visit us, their technology would be far beyond ours. It would be beyond anything we could imagine. As scientist Carl Sagan has said, it would look to us as if the creatures were performing "magic."

If you have been timed while reading this selection, enter your reading time on the chart below. Then turn to the Words-per-Minute table on page 112 and look up your reading speed (words per minute). Enter your reading speed on the graph on page 113.

READING TIME: Unit 12		
____________	:	____________
Minutes		*Seconds*

How Well Did You Read?

- *Complete the four exercises that follow. The directions for each exercise will tell you how to mark your answers.*
- *When you have finished all four exercises, use the answer key on page 108 to check your work. For each right answer, put a check mark (✓) on the line beside the box. For each wrong answer, write the correct answer on the line.*
- *Follow the directions after each exercise to find your scores.*

A FINDING THE MAIN IDEA

A good main idea statement answers two questions: it tells *who* or *what* is the subject of the story, and it answers the understood question *does what?* or *is what?* Look at the three statements below. One expresses the main idea of the story you just read. Another statement is *too broad*; it is vague and doesn't tell much about the topic of the story. The third statement is *too narrow*; it tells about only one part of the story.

Match the statements with the three answer choices below by writing the letter of each answer in the box in front of the statement it goes with.

M—Main Idea **B—Too Broad** **N—Too Narrow**

___ ☐ 1. Travel to the nearest star would take our space ships 100,000 years.

___ ☐ 2. Many people believe in intelligent life on other planets.

___ ☐ 3. Although intelligent life may exist on other planets, you are not likely, for a variety of reasons, to meet an alien here on Earth.

___ Score 15 points for a correct *M* answer.
___ Score 5 points for each correct *B* or *N* answer.

___ TOTAL SCORE: Finding the Main Idea

B RECALLING FACTS

How well do you remember the facts in the story you just read? Put an *x* in the box in front of the correct answer to each of the multiple-choice questions below.

1. People on Earth have been sending radio signals into outer space for
   ____ ☐ a. only a few years.
   ____ ☐ b. centuries.
   ____ ☐ c. most of human history.

2. Scientists believe there may be intelligent life on other planets because
   ____ ☐ a. they have received messages from aliens.
   ____ ☐ b. there are many planets that seem like ours.
   ____ ☐ c. the planets are so far away.

3. Radio waves travel
   ____ ☐ a. slower than the speed of light.
   ____ ☐ b. faster than the speed of light.
   ____ ☐ c. at the speed of light.

4. Intelligent life from other planets would probably
   ____ ☐ a. look just like humans.
   ____ ☐ b. be made up of atoms and molecules.
   ____ ☐ c. look like alligators.

5. If aliens were advanced enough to travel to Earth, they would probably think that humans were
   ____ ☐ a. not as intelligent as they were.
   ____ ☐ b. more intelligent than they were.
   ____ ☐ c. just as intelligent as they were.

Score 5 points for each correct answer.

____ TOTAL SCORE: Recalling Facts

C MAKING INFERENCES

When you use information from the text and your own experience to draw a conclusion that is not directly stated in the text, you are making an *inference*.

Below are five statements that may or may not be inferences based on the facts of the story. Write the letter *C* in the box in front of each statement that is a correct inference. Write the letter *F* in front of each faulty inference.

C—Correct Inference F—Faulty Inference

____ ☐ 1. The major reason for human efforts to make contact with other intelligent life is human curiosity.

____ ☐ 2. If alien life forms ever come to Earth, their most likely reason will be to destroy us, their rivals.

____ ☐ 3. As scientists discover more stars, more of them expect that other stars' planets will have the right conditions for intelligent life.

____ ☐ 4. If friendly aliens ever visit Earth, we could apply much of their medical knowledge to ourselves.

____ ☐ 5. Unless other intelligent beings can translate the languages used in our radio broadcasts, they will never realize Earth has intelligent life.

Score 5 points for each correct *C* or *F* answer.

____ TOTAL SCORE: Making Inferences

USING WORDS PRECISELY

Each numbered sentence below contains an underlined word or phrase from the story you have just read. Under the sentence are three definitions. One is a *synonym,* a word that means the same or almost the same thing as the underlined word: *big* and *large* are synonyms. One is an *antonym,* a word that has the opposite or nearly opposite meaning: *love* and *hate* are antonyms. One is an unrelated word; it has a completely *different* meaning than the underlined word. Match the definitions with the three answer choices by writing the letter that stands for each answer in the box in front of the definition it goes with.

S—Synonym A—Antonym D—Different

1. As proof, these people would point to UFOs (Unidentified Flying Objects).

___ ☐ a. powerful
___ ☐ b. not recognized
___ ☐ c. known

2. These signals have not had time to reach any distant planets.

___ ☐ a. far away
___ ☐ b. strange
___ ☐ c. close

3. Besides, suppose that some life form *is* that advanced.

___ ☐ a. curious
___ ☐ b. behind the times
___ ☐ c. ahead of our time

4. It's frustrating to think that we might never see the life forms that inhabit other planets.

___ ☐ a. soothing
___ ☐ b. annoying
___ ☐ c. funny

5. By the time we get these signals, the creatures who sent them might no longer exist.

___ ☐ a. listen
___ ☐ b. live
___ ☐ c. die out

___ Score 3 points for a correct *S* answer.
___ Score 1 point for each correct *A* or *D* answer.

___ TOTAL SCORE: Using Words Precisely

- *Enter the total score for each exercise in the spaces below. Add the scores to find your Critical Reading Score. Then record your Critical Reading Score on the graph on page 114.*

______	Finding the Main Idea
______	Recalling Facts
______	Making Inferences
______	Using Words Precisely
______	CRITICAL READING SCORE: Unit 12

"The excitement gave me a stomachache." "I was so nervous that my heart was racing." Such statements are often true. Your emotions—and even your thoughts—affect your stomach, heart, and other organs. We can't get inside our own bodies to see the effect of certain thoughts or emotions. But we can use modern technology to get clues. In these pictures, researchers attach special wires to patients' skin to find out what's happening inside. The wires pick up information about their involuntary actions, such as heart beat and nerve impulses. Then, using the information, the patients learn how to control some of these actions.

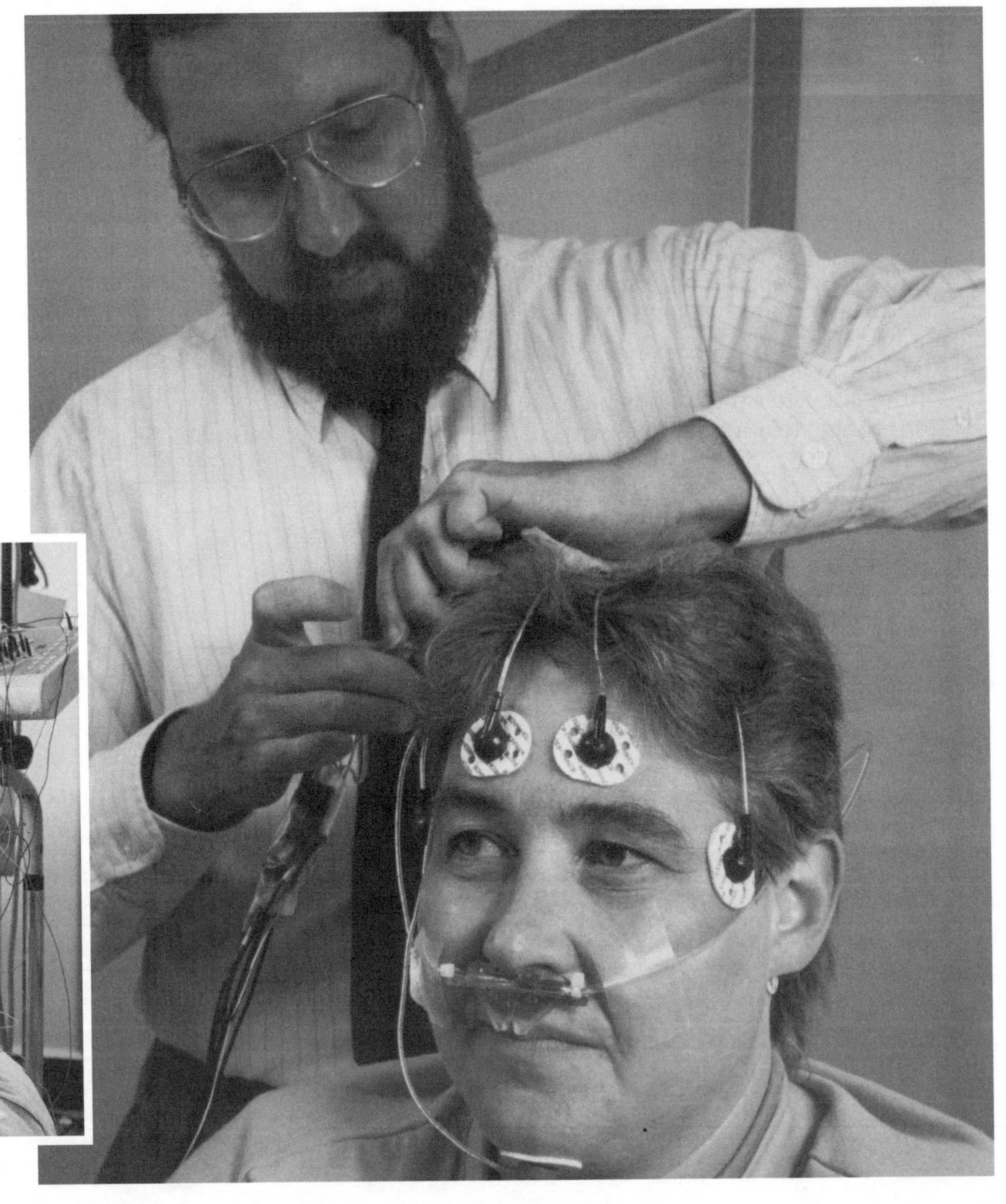

IT'S ALL IN YOUR HEAD

"You are what you eat." Have you ever been told that as you pop potato chips into your mouth? Actually, it's a pretty fair warning. Your body turns the food you eat into blood, bones, and muscles. Eat bad food and you will probably end up with a weak body. Eat good food and you will probably end up with a good body.

Everyone agrees that eating healthy foods will help make a healthy body. But now scientists are also discovering that "you are what you think." That doesn't mean you can think your way into the Olympics. And it doesn't mean that you can close your eyes and heal a broken leg. But your thoughts, or brain waves, can do some remarkable things. They can cure headaches. They can help lower blood pressure. Your thoughts can even direct messages to certain nerve cells in your body. This ability can restore the use of muscles lost through an accident or disease.

This new form of mind control is called *biofeedback*. In truth, it isn't really new. People in Asia have been using biofeedback for centuries. In the United States, however, use of biofeedback is new. It began during the 1960s. Some young people began studying the religions of Asia. They learned about biofeedback and began to practice it. These people used it to help them meditate, or focus their thinking. At first, Western scientists laughed at biofeedback. They thought it was just a fad. Slowly, however, the practice took hold. One study after another showed that it really worked. Maybe, scientists thought, biofeedback wasn't so crazy after all.

Clearly, your brain can control certain things. For example, it can command your legs to run. It can direct your hand to write a letter. And it can tell your mouth to speak. Those are called *voluntary* actions. You make up your mind to do something. But your body does other things that are very hard to control. How, for instance, can you control your heartbeat? Or the circulation of your blood? Or your body's temperature? Those bodily functions are called *involuntary*. They just happen. You have no control over them . . . or do you? The goal of biofeedback is to help you control those involuntary actions.

Here is how it works. You sit in a chair in front of a TV monitor. Special wires are taped to your head, neck, back, and fingers. The wires can detect tiny changes in your body's involuntary actions. The other ends of the wires are hooked up to a biofeedback machine. Let's say that you are trying to reduce your heart rate. Soft music begins to play. You try to relax. You might try deep breathing or you might concentrate on thinking pleasant thoughts. You might imagine yourself lying on the beach in the warm sun. The machine measures your heart rate and "feeds back" information about how well you are doing.

You can see your progress on the TV screen. A line or a series of beeps records your heart rate. By watching that line, you can learn what thoughts slow down your heart. You might find that an image of a sunny beach lowers your heart rate. But imagining that same beach bathed in moonlight does not. Slowly, you learn how to reduce your heart rate. Once this is done, you no longer need the machine. You have learned to control your body's "involuntary" actions.

But biofeedback can work even greater wonders. Since the 1970s, it has been used to help epileptics control seizures. These seizures cause epileptics to shake without control. Sometimes they faint. These attacks are caused by electrical discharges in the brain.

Using biofeedback, many epileptics have learned to control their seizures. They discover what feelings or situations trigger seizures. They learn to avoid those feelings. In a sense, then, biofeedback shows epileptics how to "rewire" their brains.

Biofeedback can even help badly hurt people. In the 1980s, a car accident left Tammy DeMichael with a broken neck and a crushed spinal cord. She had no feeling in her arms or legs. Traditional medicine did nothing to help her. It looked as if she would have to spend the rest of her life in a wheelchair. Luckily, she still had a few good nerves reaching her arm muscles. They were not enough to let DeMichael move her arm, but they gave her some hope.

Could her brain be taught to use those remaining nerves? Bernard Brucker, her doctor, thought so. He hooked DeMichael up to a biofeedback machine. The TV monitor showed a blue line. That line represented impulses moving from her brain through her spine to her arm muscles. DeMichael concentrated as hard as she could. Slowly, she got the line on the TV screen to move up. Still, her arm didn't move. But the line showed that more impulses were reaching her arm. One day, DeMichael got the arm to move. Everyone in the room cheered. She did the same thing with her legs. It took several years, but biofeedback worked. DeMichael learned to walk using just a cane.

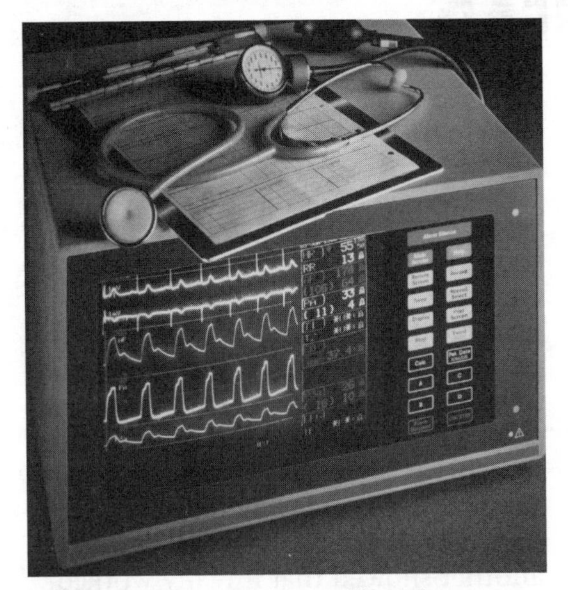

Biofeedback machines, like the one pictured, measure reactions such as heart rate, blood pressure, and nerve impulses. By changing their thoughts and feelings, patients can change what they see on the machine's screens.

Western scientists no longer scoff at biofeedback. They see it as a hot new medical tool. It has been used to teach some children how to pay better attention in school. Astronauts use it to fight motion sickness during space travel. It has even been used to treat certain forms of cancer. Clearly, biofeedback is here to stay. Bernard Brucker put it this way. "Biofeedback," he says, "has opened up a whole new era in human learning."

If you have been timed while reading this selection, enter your reading time on the chart below. Then turn to the Words-per-Minute table on page 112 and look up your reading speed (words per minute). Enter your reading speed on the graph on page 113.

READING TIME: Unit 13		
____________	:	____________
Minutes		*Seconds*

How Well Did You Read?

- *Complete the four exercises that follow. The directions for each exercise will tell you how to mark your answers.*
- *When you have finished all four exercises, use the answer key on page 108 to check your work. For each right answer, put a check mark (✓) on the line beside the box. For each wrong answer, write the correct answer on the line.*
- *Follow the directions after each exercise to find your scores.*

A FINDING THE MAIN IDEA

A good main idea statement answers two questions: it tells *who* or *what* is the subject of the story, and it answers the understood question *does what?* or *is what?* Look at the three statements below. One expresses the main idea of the story you just read. Another statement is *too broad*; it is vague and doesn't tell much about the topic of the story. The third statement is *too narrow*; it tells about only one part of the story.

Match the statements with the three answer choices below by writing the letter of each answer in the box in front of the statement it goes with.

M—Main Idea **B—Too Broad** **N—Too Narrow**

___ ☐ 1. Until a few years ago, biofeedback was not taken seriously by Western scientists.

___ ☐ 2. Biofeedback is helping more people take control of their bodies.

___ ☐ 3. Scientists are eager to find new ways to treat diseases and other physical problems.

___ Score 15 points for a correct *M* answer.
___ Score 5 points for each correct *B* or *N* answer.

___ TOTAL SCORE: Finding the Main Idea

B RECALLING FACTS

How well do you remember the facts in the story you just read? Put an *x* in the box in front of the correct answer to each of the multiple-choice questions below.

1. When Western scientists first heard about biofeedback, they thought it
   - ___ ☐ a. was just a fad.
   - ___ ☐ b. was useful only for young people.
   - ___ ☐ c. made a lot of sense.

2. Body actions that are hard to control are called
   - ___ ☐ a. electrical.
   - ___ ☐ b. voluntary.
   - ___ ☐ c. involuntary.

3. In biofeedback, wires attached to your body
   - ___ ☐ a. control body functions.
   - ___ ☐ b. detect tiny changes.
   - ___ ☐ c. give you small electrical charges.

4. A car accident left Tammy DeMichael with
   - ___ ☐ a. no sense of balance.
   - ___ ☐ b. a broken neck and a crushed spinal cord.
   - ___ ☐ c. no arms or legs.

5. Epileptic seizures are caused by
   - ___ ☐ a. electrical discharges in the brain.
   - ___ ☐ b. improper diet.
   - ___ ☐ c. too many chemicals in the brain.

Score 5 points for each correct answer.

___ TOTAL SCORE: Recalling Facts

C MAKING INFERENCES

When you use information from the text and your own experience to draw a conclusion that is not directly stated in the text, you are making an *inference*.

Below are five statements that may or may *not* be inferences based on the facts of the story. Write the letter *C* in the box in front of each statement that is a correct inference. Write the letter *F* in front of each faulty inference.

C—Correct Inference **F—Faulty Inference**

___ ☐ 1. Soon, diseases will be wiped out by biofeedback.

___ ☐ 2. Some Asian religions make use of the methods that are part of biofeedback.

___ ☐ 3. If you are always upset and nervous, your body systems will be affected.

___ ☐ 4. In most cases, biofeedback works better than medicine.

___ ☐ 5. Western scientists never trust any methods of health care that come from religion.

Score 5 points for each correct *C* or *F* answer.

___ TOTAL SCORE: Making Inferences

D USING WORDS PRECISELY

Each numbered sentence below contains an underlined word or phrase from the story you have just read. Under the sentence are three definitions. One is a *synonym*, a word that means the same or almost the same thing as the underlined word: *big* and *large* are synonyms. One is an *antonym*, a word that has the opposite or nearly opposite meaning: *love* and *hate* are antonyms. One is an unrelated word; it has a completely *different* meaning than the underlined word. Match the definitions with the three answer choices by writing the letter that stands for each answer in the box in front of the definition it goes with.

S—Synonym A—Antonym D—Different

1. But your thoughts, or brain waves, can do some remarkable things.

___ ☐ a. amazing
___ ☐ b. nice
___ ☐ c. normal

2. The wires can detect tiny changes in your body's involuntary actions.

___ ☐ a. ignore
___ ☐ b. cause
___ ☐ c. discover

3. Slowly, you learn how to reduce your heart rate.

___ ☐ a. increase
___ ☐ b. lower
___ ☐ c. pay attention to

4. Your thoughts can even restore the use of muscles lost through an accident or disease.

___ ☐ a. put back in working order
___ ☐ b. damage
___ ☐ c. display

5. Western scientists no longer scoff at biofeedback.

___ ☐ a. honor
___ ☐ b. make fun of
___ ☐ c. point to

___ Score 3 points for a correct *S* answer.
___ Score 1 point for each correct *A* or *D* answer.

___ TOTAL SCORE: Using Words Precisely

- *Enter the total score for each exercise in the spaces below. Add the scores to find your Critical Reading Score. Then record your Critical Reading Score on the graph on page 114.*

______	Finding the Main Idea
______	Recalling Facts
______	Making Inferences
______	Using Words Precisely
______	CRITICAL READING SCORE: Unit 13

No, this is not an Egyptian mummy wrapped in foil. Long ago, the Egyptians preserved a dead person's body as a mummy just in case the person's spirit ever needed its body again. Today, some people preserve dead bodies for a similar reason. Instead of making mummies, however, they use a method of freezing called cryonics. *In the future, whatever disease killed the person may be curable. Then the person's body may be thawed out and cured. Cryonics gives people hope that they will live again.*

CRYONICS: DEATH ON ICE

Like most people, Dick Clair wanted to live a long life. He didn't quite make it, however. At the age of fifty-seven, he grew ill and died. But Clair, a TV comedy writer, was determined to have the last laugh. So he arranged to have his body frozen instead of buried in the ground. Someday, Clair hoped, scientists would find a cure for the disease that killed him. Then doctors could thaw out his body and cure him. Once revived, Clair could start writing TV comedy shows again.

Like a growing number of people, Clair believed in cryonics. *Cryonics* is the practice of freezing a body at the moment of death. Cryonic suspension was first done in 1968. Since then, many people have been frozen. Hundreds more have signed up for future freezing. Some even plan to have their favorite dog or cat frozen with them. That way, owner and pet may be reunited when a cure is found.

Cyronics takes careful planning. As soon as someone dies, oxygen must be pumped into the body. That keeps the body tissues from decaying. The body must also be packed in ice to keep it cool. Then the blood is drained and replaced with a special fluid. Next, the body is wrapped in plastic and zipped into a sleeping bag. Finally, the bag is put into a nine-foot metal tube. A special cooling gas, called *liquid nitrogen,* is pumped into the tube. Slowly, the temperature drops all the way to -321°F. At such a super-low temperature, a body will last almost forever. The decay that would take one second at room temperature would take 30 trillion years at -321°F! So people such as Clair can be kept on ice for a very long time.

Cyronics is not a proven science. There are still plenty of bugs to be worked out. One of the biggest hurdles is the freezing process itself. Parts of a human body can be preserved for a short time. A heart, for example, can be saved for several hours. This has made human heart transplants possible. But saving a *whole body* for a long time is much harder.

Freezing will preserve it. But freezing also tends to destroy it. The human body is made up of cells surrounded by water. As soon as the temperature drops below 32°F, that water starts to expand. It forms ice crystals. That causes tearing and damages body tissue. All bodies in cryonic suspension have suffered tissue damage. There is no guarantee that scientists will ever figure out a way to repair the damage.

Let's assume, however, that they do. Let's also assume that scientists solve all the other technical problems. Moral questions would still remain. Who should get frozen? Cryonic suspension is not cheap. At present, it costs more than $100,000. Some people opt for a cheaper solution. They pay $35,000 each to have just their heads frozen. They are gambling that scientists will someday be able to attach each head to a brand new body. Still, freezing any part of the body takes big bucks. Does that mean only rich people should get a second chance at life?

Surely, not everyone can be frozen. The world already has too many people. What would happen if lots of "dead" people came back to life? The world could not support them all. So someone would have to decide who gets frozen and who doesn't.

Should the young be favored over the old? Should law-abiding people be picked over criminals? Would a musician be selected ahead of a street sweeper?

There are other problems as well. Let's suppose a wife dies and is cryonically suspended. Is her widower free to remarry? What happens to her money and property? If she is thawed out, can she move back into her old house? Can she demand her old job back? How would she talk to her kids if she's now twenty or thirty years *younger* than they are? Suppose she doesn't come back for a thousand years. What would she do in this strange new world? How would she make a living? Who would be her friends?

The questions go on and on. Still, people cling to the hope that they can someday live again. In 1993, *Omni* magazine held an essay contest. The winner got a free cryonic suspension. *Omni* received hundreds of essays. People gave lots of good reasons for wanting to live again. Some wrote that they were excited about the distant future and wanted to see it for themselves. Others wanted to carry knowledge of today's world to the people of the future. Still others felt that they had missed out on things in this life and wanted a second chance. One young reader—the winner—had been injured in a car accident. He wanted to come back "healthy and healed."

Cryonics may prove to be a false hope. All the frozen dead bodies may be just that—frozen and dead. But cryonics brings hope to those who believe in it. Even if it doesn't work out, what have they lost besides the money? As Dick Clair once said, "To me, [cryonics] is a way to stay alive."

If you have been timed while reading this selection, enter your reading time on the chart below. Then turn to the Words-per-Minute table on page 112 and look up your reading speed (words per minute). Enter your reading speed on the graph on page 113.

READING TIME: Unit 14		
____________	:	____________
Minutes		*Seconds*

How Well Did You Read?

- *Complete the four exercises that follow. The directions for each exercise will tell you how to mark your answers.*
- *When you have finished all four exercises, use the answer key on page 108 to check your work. For each right answer, put a check mark (✓) on the line beside the box. For each wrong answer, write the correct answer on the line.*
- *Follow the directions after each exercise to find your scores.*

A FINDING THE MAIN IDEA

A good main idea statement answers two questions: it tells *who* or *what* is the subject of the story, and it answers the understood question *does what?* or *is what?* Look at the three statements below. One expresses the main idea of the story you just read. Another statement is *too broad*; it is vague and doesn't tell much about the topic of the story. The third statement is *too narrow*; it tells about only one part of the story.

Match the statements with the three answer choices below by writing the letter of each answer in the box in front of the statement it goes with.

M—Main Idea B—Too Broad N—Too Narrow

___ ☐ 1. Cryonics is the art of freezing a body at the moment of death.

___ ☐ 2. Some people are paying $35,000 to have just their heads frozen after death.

___ ☐ 3. Cryonics offers hope for renewed life, but also poses problems that have not yet been worked out.

___ Score 15 points for a correct *M* answer.
___ Score 5 points for each correct *B* or *N* answer.

___ TOTAL SCORE: Finding the Main Idea

B RECALLING FACTS

How well do you remember the facts in the story you just read? Put an *x* in the box in front of the correct answer to each of the multiple-choice questions below.

1. Cryonics suspension was first done in
   - ____ ☐ a. 1995.
   - ____ ☐ b. 1968.
   - ____ ☐ c. 1924.

2. A body ready to be frozen is put into a metal tube, which is then filled with
   - ____ ☐ a. oxygen.
   - ____ ☐ b. ice.
   - ____ ☐ c. liquid nitrogen.

3. When the temperature drops below 32° F, water
   - ____ ☐ a. starts to expand.
   - ____ ☐ b. starts to shrink.
   - ____ ☐ c. starts to boil.

4. Ice crystals in body cells cause
   - ____ ☐ a. tissue damage.
   - ____ ☐ b. the body to stop aging.
   - ____ ☐ c. moral problems.

5. Bodies kept at -321°F will resist decay
   - ____ ☐ a. for about fifty years.
   - ____ ☐ b. for about one thousand years.
   - ____ ☐ c. almost forever.

Score 5 points for each correct answer.

____ TOTAL SCORE: Recalling Facts

C MAKING INFERENCES

When you use information from the text and your own experience to draw a conclusion that is not directly stated in the text, you are making an *inference*.

Below are five statements that may or may *not* be inferences based on the facts of the story. Write the letter *C* in the box in front of each statement that is a correct inference. Write the letter *F* in front of each faulty inference.

C—Correct Inference **F—Faulty Inference**

____ ☐ 1. People who spend their money on cryonic suspension have faith in the future.

____ ☐ 2. Most people who have undergone cryonic suspension were wealthy when they were alive.

____ ☐ 3. Someday there will be a worldwide organization that will decide who will be frozen.

____ ☐ 4. If the temperature of bodies that are frozen rises too high, the bodies will begin to decay.

____ ☐ 5. Most people who die from now on will have their bodies frozen so they can live again later.

Score 5 points for each correct *C* or *F* answer.

____ TOTAL SCORE: Making Inferences

D USING WORDS PRECISELY

Each numbered sentence below contains an underlined word or phrase from the story you have just read. Under the sentence are three definitions. One is a *synonym*, a word that means the same or almost the same thing as the underlined word: *big* and *large* are synonyms. One is an *antonym*, a word that has the opposite or nearly opposite meaning: *love* and *hate* are antonyms. One is an unrelated word; it has a completely *different* meaning than the underlined word. Match the definitions with the three answer choices by writing the letter that stands for each answer in the box in front of the definition it goes with.

S—Synonym A—Antonym D—Different

1. Once <u>revived</u>, Clair could start writing TV comedy shows again.

___ ☐ a. restored to life
___ ☐ b. frozen
___ ☐ c. put to death

2. That way, owner and pet may be <u>reunited</u> when a cure is found.

___ ☐ a. remembered
___ ☐ b. brought back together
___ ☐ c. separated again

3. One of the biggest <u>hurdles</u> is the freezing process itself.

___ ☐ a. signals
___ ☐ b. helps
___ ☐ c. obstacles

4. The <u>decay</u> that would take one second at room temperature would take 30 trillion years at -321°F!

___ ☐ a. building up of the body
___ ☐ b. breakdown of the body
___ ☐ c. change in the body

5. That causes tearing and <u>damages</u> body tissues.

___ ☐ a. improves
___ ☐ b. affects
___ ☐ c. harms

___ Score 3 points for a correct *S* answer.
___ Score 1 point for each correct *A* or *D* answer.

___ TOTAL SCORE: Using Words Precisely

• *Enter the total score for each exercise in the spaces below. Add the scores to find your Critical Reading Score. Then record your Critical Reading Score on the graph on page 114.*

______ Finding the Main Idea
______ Recalling Facts
______ Making Inferences
______ Using Words Precisely

______ CRITICAL READING SCORE: Unit 14

Although it looks fragile, this monarch butterfly is a hardy traveler. Each fall, it flies about 2,000 miles to get from New England to Mexico. Why? To escape the cold northern winter. The monarch is only one of thousands of animals—birds, fish, insects, and other land and sea creatures—that migrate. Not all migrate just for better weather. Many return to their birthplaces to lay their eggs or bear young. And the monarch's trip is not the most remarkable journey. Many travel faster or farther. What scientists are still trying to find out is how these animals find their way home over thousands of miles of unfamiliar territory.

ANIMAL MIGRATION

"There's no place like home."

That's what Dorothy said at the end of the movie *The Wizard of Oz*. Apparently, many animals agree with her. Consider the puffin. This Arctic bird always knows how to find its home. A puffin roams hundreds of miles during the fall and winter, but every spring it returns home. It flies back to the exact spot where it built its nest. It doesn't matter how much snow hides the nest. It doesn't matter if rocks have tumbled onto the area. It doesn't even matter if the nest has been completely destroyed. To a puffin, home will always be home. In Iceland one year, puffin nests were destroyed when a volcano erupted. Hot lava covered the nests. Still, puffins insisted on returning to the spot. The birds died when they tried to land on the steaming lava.

Like puffins, greater yellowlegs love to travel. In fact, every year these birds fly from northern Canada to the tip of South America. That's a round trip of eighteen thousand miles! Despite the long trip, the greater yellowlegs are incredibly prompt. Each year, without fail, they make it back to Canada in time to hatch their eggs between May 26 and May 29.

The olive ridley sea turtles spend much of the year swimming in the ocean. Yet they return to the exact same beach each October. Females drag their eighty-pound bodies slowly across the coarse black sand. Soon, about thirty thousand turtles cover the beach at Ostional, Costa Rica. The shells of the turtles make the beach glisten like a cobblestone street. In all, the turtles lay about 20 million eggs in the sand. Then they swim back out to sea.

These animals are migrants. They travel without the aid of a travel agent or road signs or maps. They just go. And some of their stories are really remarkable. Think about the blackpoll warbler. This bird weighs less than one ounce. Yet every fall, it leaves Alaska and flies across Canada to Nova Scotia. As winter approaches, the warbler takes off again. This time it flies about 2,400 miles to South America. The tiny warbler makes this flight *nonstop*!

Even a butterfly can go a long way. Every fall the monarch butterfly flutters two thousand miles from New England to Mexico. It covers up to eighty miles a day during this trip. The butterfly knows where it's going. It heads straight for a certain grove of trees on a Mexican mountainside.

The all-time distance champ is the arctic tern. Each year this bird flies about 24,000 miles. It travels from near the North Pole to Antarctica and back.

Why do these animals migrate such great distances? Most travel in search of food or a good place to breed. That often means they follow the warm sun. So in the Northern Hemisphere, they head south in the fall. They move back north in the spring.

Many animals are born with an irresistible urge to migrate. Take the case of the green sea turtle. It lives along the coast of Brazil. But this turtle does not lay its eggs in Brazil—or anywhere else in South America. Instead, it heads for the sandy beaches of Ascension Island in the South Atlantic. That's about 1,300 miles away. It takes the poor turtle eight weeks to paddle there. Yet every year the green sea turtle makes this trip. After laying its eggs, the turtle swims back to Brazil.

Why does the green sea turtle pick an island so far away? After all, there are plenty of beaches closer to home. Archie Carr, a zoologist, thinks he knows why. It's tradition, a very old tradition. The green sea turtle has been going to Ascension Island for at least 80 million years. At first, Carr believes, the island was just four miles from Brazil. Ancient turtles found it a safe and convenient spot to lay their eggs. Over the ages, the island slowly drifted away. It is now halfway across the Atlantic Ocean. But the green sea turtles didn't notice the change. That's because the island probably drifted only a few inches in an average turtle's life.

All migrating animals seem to know just when it's time to pick up and move. How do they do it? Scientists have looked at several factors. It could be that falling temperatures tell them it's time to go. It could be that they take their cue from changing winds or the position of the stars. Zoologist William Rowan thinks the key is the changing number of daylight hours. As the days grow shorter, animals sense that winter is coming. So they head south. When the number of daylight hours grows longer, they know it is time to return north.

Still, a big question remains. How do these animals navigate? How do they know where they are going? How can the puffin find a nest that is buried under snow? How can turtles find the same beach across a thousand miles of open sea? And how is it that monarch butterflies never miss that grove of trees?

Much remains a mystery. But scientists are coming up with a few ideas. One suggestion is that birds use the sun to navigate. That's fine on sunny days, but they don't get lost when it's cloudy, either. So there must be something else to guide them. Scientist William Keeton thinks the answer lies with magnets. He points out that the North Pole is a kind of magnet. It pulls things—such as the needle of a compass—toward it. If birds could feel this pull, they could figure out which way is north. In fact, scientists have shown that homing pigeons do have special crystals, called *magnetite,* in their heads. These tiny crystals act as magnets. So in effect, a homing pigeon has a built-in compass that allows it to find its way even on cloudy days.

Animals use other clues to help them navigate. Birds that fly at night use the stars to guide them. They also use landmarks. Birds have the best eyesight of any animal in the world. So they can spot a familiar river or shoreline from high in the sky. Certain fish navigate by odor. Salmon, for example, find their birthplace by following smells in the water. Migratory animals may also use sounds and lights that humans cannot hear or see. Animal migration is one of nature's great wonders. And although scientists don't always understand it, the important thing is that the animals themselves know what they are doing!

If you have been timed while reading this selection, enter your reading time on the chart below. Then turn to the Words-per-Minute table on page 112 and look up your reading speed (words per minute). Enter your reading speed on the graph on page 113.

READING TIME: Unit 15		
______	:	______
Minutes		*Seconds*

How Well Did You Read?

- *Complete the four exercises that follow. The directions for each exercise will tell you how to mark your answers.*
- *When you have finished all four exercises, use the answer key on page 108 to check your work. For each right answer, put a check mark (✓) on the line beside the box. For each wrong answer, write the correct answer on the line.*
- *Follow the directions after each exercise to find your scores.*

A FINDING THE MAIN IDEA

A good main idea statement answers two questions: it tells *who* or *what* is the subject of the story, and it answers the understood question *does what?* or *is what?* Look at the three statements below. One expresses the main idea of the story you just read. Another statement is *too broad*; it is vague and doesn't tell much about the topic of the story. The third statement is *too narrow*; it tells about only one part of the story.

Match the statements with the three answer choices below by writing the letter of each answer in the box in front of the statement it goes with.

M—Main Idea B—Too Broad N—Too Narrow

___ ☐ 1. Many animals migrate, and scientists have a variety of theories about how and why they do it.

___ ☐ 2. Migration seems to be a common occurrence in the animal kingdom.

___ ☐ 3. Olive ridley turtles return to the same beach each October to lay their eggs.

___ Score 15 points for a correct *M* answer.
___ Score 5 points for each correct *B* or *N* answer.

___ TOTAL SCORE: Finding the Main Idea

B RECALLING FACTS

How well do you remember the facts in the story you just read? Put an *x* in the box in front of the correct answer to each of the multiple-choice questions below.

1. When hot lava covered their nests, some puffins
   - ____ ☐ a. built new nests in a different place.
   - ____ ☐ b. did not migrate that year.
   - ____ ☐ c. died because they tried to land on their old nests anyway.

2. One amazing fact about the blackpoll warbler is
   - ____ ☐ a. it flies 2,400 miles nonstop each fall.
   - ____ ☐ b. it lays its eggs on a beach in Costa Rica.
   - ____ ☐ c. its call is very loud for such a small bird.

3. The tiny arctic tern travels about
   - ____ ☐ a. 2,400 miles.
   - ____ ☐ b. 24,000 miles.
   - ____ ☐ c. 100,000 miles.

4. Most animals migrate because they
   - ____ ☐ a. are being hunted by other animals.
   - ____ ☐ b. are looking for food and a place to breed.
   - ____ ☐ c. have nothing better to do.

5. Scientists believe that a homing pigeon's head
   - ____ ☐ a. is always turned north during the day.
   - ____ ☐ b. holds a map of every place the pigeon has flown.
   - ____ ☐ c. has crystals that act like magnets.

Score 5 points for each correct answer.

____ TOTAL SCORE: Recalling Facts

C MAKING INFERENCES

When you use information from the text and your own experience to draw a conclusion that is not directly stated in the text, you are making an *inference*.

Below are five statements that may or may *not* be inferences based on the facts of the story. Write the letter *C* in the box in front of each statement that is a correct inference. Write the letter *F* in front of each faulty inference.

C—Correct Inference F—Faulty Inference

____ ☐ 1. Olive ridley turtles care for their young until they are grown.

____ ☐ 2. The blackpoll warbler can fly for days without stopping to sleep.

____ ☐ 3. In the Southern Hemisphere, birds probably fly north in April.

____ ☐ 4. Land forms such as islands always stay in the same place on the surface of the earth.

____ ☐ 5. The average dog could probably spot a bird before the bird saw it.

Score 5 points for each correct *C* or *F* answer.

____ TOTAL SCORE: Making Inferences

USING WORDS PRECISELY

Each numbered sentence below contains an underlined word or phrase from the story you have just read. Under the sentence are three definitions. One is a *synonym*, a word that means the same or almost the same thing as the underlined word: *big* and *large* are synonyms. One is an *antonym*, a word that has the opposite or nearly opposite meaning: *love* and *hate* are antonyms. One is an unrelated word; it has a completely *different* meaning than the underlined word. Match the definitions with the three answer choices by writing the letter that stands for each answer in the box in front of the definition it goes with.

S—Synonym A—Antonym D—Different

1. Despite the long trip, the greater yellowlegs are incredibly prompt.

___ ☐ a. late
___ ☐ b. active
___ ☐ c. on time

2. These animals are migrants.

___ ☐ a. ones who travel
___ ☐ b. ones who stay in one place
___ ☐ c. ones who remember

3. Females drag their eighty-pound bodies slowly across the coarse black sand.

___ ☐ a. rough
___ ☐ b. hot
___ ☐ c. fine

4. As winter approaches, the warbler takes off again.

___ ☐ a. goes away
___ ☐ b. comes closer
___ ☐ c. gets colder

5. Many animals are born with an irresistible urge to migrate.

___ ☐ a. natural
___ ☐ b. easy to resist
___ ☐ c. impossible to resist

___ Score 3 points for a correct *S* answer.
___ Score 1 point for each correct *A* or *D* answer.

___ TOTAL SCORE: Using Words Precisely

- *Enter the total score for each exercise in the spaces below. Add the scores to find your Critical Reading Score. Then record your Critical Reading Score on the graph on page 114.*

_____ Finding the Main Idea
_____ Recalling Facts
_____ Making Inferences
_____ Using Words Precisely

_____ CRITICAL READING SCORE: Unit 15

Answer Key

1 The Mysterious Life of Twins
Pages 10–15

A. Finding the Main Idea
1. M 2. N 3. B

B. Recalling Facts
1. a 2. b 3. c 4. a 5. c

C. Making Inferences
1. F 2. C 3. F 4. F 5. C

D. Using Words Precisely
1. a. S b. A c. D
2. a. A b. D c. S
3. a. A b. S c. D
4. a. D b. A c. S
5. a. A b. S c. D

2 Is the Earth Alive?
Pages 16–21

A. Finding the Main Idea
1. B 2. N 3. M

B. Recalling Facts
1. c 2. b 3. c 4. a 5. b

C. Making Inferences
1. F 2. C 3. F 4. C 5. F

D. Using Words Precisely
1. a. A b. S c. D
2. a. S b. D c. A
3. a. A b. D c. S
4. a. S b. A c. D
5. a. D b. S c. A

3 Worms, Worms, Worms
Pages 22–27

A. Finding the Main Idea
1. B 2. M 3. N

B. Recalling Facts
1. b 2. b 3. b 4. c 5. a

C. Making Inferences
1. C 2. F 3. C 4. F 5. F

D. Using Words Precisely
1. a. A b. S c. D
2. a. S b. D c. A
3. a. A b. S c. D
4. a. A b. D c. S
5. a. D b. A c. S

4 Fire Storms
Pages 28–33

A. Finding the Main Idea
1. B 2. N 3. M

B. Recalling Facts
1. b 2. c 3. a 4. a 5. c

C. Making Inferences
1. C 2. C 3. F 4. F 5. C

D. Using Words Precisely
1. a. S b. D c. A
2. a. A b. D c. S
3. a. S b. A c. D
4. a. A b. S c. D
5. a. S b. D c. A

5 Dowsing: Fact or Fiction?
Pages 34–39

A. Finding the Main Idea
1. N 2. B 3. M

B. Recalling Facts
1. c 2. a 3. c 4. b 5. a

C. Making Inferences
1. C 2. F 3. C 4. F 5. F

D. Using Words Precisely
1. a. A b. D c. S
2. a. S b. A c. D
3. a. D b. S c. A
4. a. A b. D c. S
5. a. A b. S c. D

6 Traveling Through Time
Pages 42–47

A. Finding the Main Idea
1. M 2. B 3. N

B. Recalling Facts
1. c 2. a 3. b 4. c 5. b

C. Making Inferences
1. F 2. C 3. F 4. C 5. F

D. Using Words Precisely
1. a. A b. S c. D
2. a. S b. A c. D
3. a. D b. A c. S
4. a. S b. A c. D
5. a. D b. S c. A

7 Can We Bring Back the Dinosaurs?
Pages 48–53

A. Finding the Main Idea
1. M 2. N 3. B

B. Recalling Facts
1. b 2. a 3. c 4. b 5. c

C. Making Inferences
1. C 2. F 3. C 4. F 5. F

D. Using Words Precisely
1. a. A b. S c. D
2. a. D b. S c. A
3. a. D b. A c. S
4. a. S b. D c. A
5. a. D b. A c. S

8 The Healing Power of Maggots
Pages 54–59

A. Finding the Main Idea
1. M 2. N 3. B

B. Recalling Facts
1. a 2. b 3. a 4. a 5. a

C. Making Inferences
1. F 2. C 3. F 4. C 5. C

D. Using Words Precisely
1. a. A b. D c. S
2. a. S b. A c. D
3. a. D b. S c. A
4. a. D b. A c. S
5. a. S b. D c. A

9 Psychics Who Solve Crimes
Pages 60–65

A. Finding the Main Idea
1. B 2. M 3. N

B. Recalling Facts
1. a 2. c 3. b 4. a 5. c

C. Making Inferences
1. F 2. C 3. F 4. C 5. C

D. Using Words Precisely
1. a. D b. A c. S
2. a. S b. D c. A
3. a. A b. S c. D
4. a. A b. D c. S
5. a. D b. S c. A

10 Mummies
Pages 66–71

A. Finding the Main Idea
1. B 2. M 3. N

B. Recalling Facts
1. b 2. c 3. c 4. b 5. a

C. Making Inferences
1. C 2. C 3. F 4. C 5. F

D. Using Words Precisely
1. a. A b. D c. S
2. a. S b. D c. A
3. a. D b. S c. A
4. a. A b. S c. D
5. a. S b. D c. A

11 Near-Death Experiences
Pages 74–79

A. Finding the Main Idea
  1. **N** 2. **B** 3. **M**
B. Recalling Facts
  1. **a** 2. **b** 3. **c** 4. **b** 5. **a**
C. Making Inferences
  1. **F** 2. **F** 3. **C** 4. **C** 5. **F**
D. Using Words Precisely
  1. a. **D** b. **S** c. **A**
  2. a. **D** b. **A** c. **S**
  3. a. **S** b. **D** c. **A**
  4. a. **A** b. **S** c. **D**
  5. a. **A** b. **D** c. **S**

12 Is Anyone Out There?
Pages 80–85

A. Finding the Main Idea
  1. **N** 2. **B** 3. **M**
B. Recalling Facts
  1. **a** 2. **b** 3. **c** 4. **b** 5. **a**
C. Making Inferences
  1. **C** 2. **F** 3. **C** 4. **F** 5. **F**
D. Using Words Precisely
  1. a. **D** b. **S** c. **A**
  2. a. **S** b. **D** c. **A**
  3. a. **D** b. **A** c. **S**
  4. a. **A** b. **S** c. **D**
  5. a. **D** b. **S** c. **A**

13 It's All in Your Head
Pages 86–91

A. Finding the Main Idea
  1. **N** 2. **M** 3. **B**
B. Recalling Facts
  1. **a** 2. **c** 3. **b** 4. **b** 5. **a**
C. Making Inferences
  1. **F** 2. **C** 3. **C** 4. **F** 5. **F**
D. Using Words Precisely
  1. a. **S** b. **D** c. **A**
  2. a. **A** b. **D** c. **S**
  3. a. **A** b. **S** c. **D**
  4. a. **S** b. **A** c. **D**
  5. a. **A** b. **S** c. **D**

14 Cryonics: Death on Ice
Pages 92–97

A. Finding the Main Idea
  1. **B** 2. **N** 3. **M**
B. Recalling Facts
  1. **b** 2. **c** 3. **a** 4. **a** 5. **c**
C. Making Inferences
  1. **C** 2. **C** 3. **F** 4. **C** 5. **F**
D. Using Words Precisely
  1. a. **S** b. **D** c. **A**
  2. a. **D** b. **S** c. **A**
  3. a. **D** b. **A** c. **S**
  4. a. **A** b. **S** c. **D**
  5. a. **A** b. **D** c. **S**

15 Animal Migration
Pages 98–103

A. Finding the Main Idea
  1. **M** 2. **B** 3. **N**
B. Recalling Facts
  1. **c** 2. **a** 3. **b** 4. **b** 5. **c**
C. Making Inferences
  1. **F** 2. **C** 3. **C** 4. **F** 5. **F**
D. Using Words Precisely
  1. a. **A** b. **D** c. **S**
  2. a. **S** b. **A** c. **D**
  3. a. **S** b. **D** c. **A**
  4. a. **A** b. **S** c. **D**
  5. a. **D** b. **A** c. **S**

Words-per-Minute Tables and Progress Graphs

WORDS PER MINUTE

GROUP ONE

Unit ➤	Sample	1	2	3	4	5	
No. of Words ➤	636	858	915	919	908	851	
1:30	424	572	610	613	605	567	90
1:40	382	515	549	551	545	511	100
1:50	347	468	499	501	495	464	110
2:00	318	429	458	460	454	426	120
2:10	294	396	422	424	419	393	130
2:20	273	368	392	394	389	365	140
2:30	254	343	366	368	363	340	150
2:40	239	322	343	345	341	319	160
2:50	224	303	323	324	320	300	170
3:00	212	286	305	306	303	284	180
3:10	201	271	289	290	287	269	190
3:20	191	257	275	276	272	255	200
3:30	182	245	261	263	259	243	210
3:40	173	234	250	251	248	232	220
3:50	166	224	239	240	237	222	230
4:00	159	215	229	230	227	213	240
4:10	153	206	220	221	218	204	250
4:20	147	198	211	212	210	196	260
4:30	141	191	203	204	202	189	270
4:40	136	184	196	197	195	182	280
4:50	132	178	189	190	188	176	290
5:00	127	172	183	184	182	170	300
5:10	123	166	177	178	176	165	310
5:20	119	161	172	172	170	160	320
5:30	116	156	166	167	165	155	330
5:40	112	151	161	162	160	150	340
5:50	109	147	157	158	156	146	350
6:00	106	143	153	153	151	142	360
6:10	103	139	148	149	147	138	370
6:20	100	135	144	145	143	134	380
6:30	98	132	141	141	140	131	390
6:40	95	129	137	138	136	128	400
6:50	93	126	134	134	133	125	410
7:00	91	123	131	131	130	122	420
7:10	89	120	128	128	127	119	430
7:20	87	117	125	125	124	116	440
7:30	85	114	122	123	121	113	450
7:40	83	112	119	120	118	111	460
7:50	81	110	117	117	116	109	470
8:00	80	107	114	115	114	106	480

Minutes and Seconds ➤   ◀ *Seconds*

GROUP TWO

Unit ➤	6	7	8	9	10	
No. of Words ➤	1053	916	854	1045	1008	
1:30	702	611	569	697	672	**90**
1:40	632	550	512	627	605	**100**
1:50	574	500	466	570	550	**110**
2:00	527	458	427	523	504	**120**
2:10	486	423	394	482	465	**130**
2:20	451	393	366	448	432	**140**
2:30	421	366	342	418	403	**150**
2:40	395	344	320	392	378	**160**
2:50	372	323	301	369	356	**170**
3:00	351	305	285	348	336	**180**
3:10	333	289	270	330	318	**190**
3:20	316	275	256	314	302	**200**
3:30	301	262	244	299	288	**210**
3:40	287	250	233	285	275	**220**
3:50	275	239	223	273	263	**230**
4:00	263	229	214	261	252	**240**
4:10	253	220	205	251	242	**250**
4:20	243	211	197	241	233	**260**
4:30	234	204	190	232	224	**270**
4:40	226	196	183	224	216	**280**
4:50	218	190	177	216	209	**290**
5:00	211	183	171	209	202	**300**
5:10	204	177	165	202	195	**310**
5:20	197	172	160	196	189	**320**
5:30	191	167	155	190	183	**330**
5:40	186	162	151	184	178	**340**
5:50	181	157	146	179	173	**350**
6:00	176	153	142	174	168	**360**
6:10	171	149	138	169	163	**370**
6:20	166	145	135	165	159	**380**
6:30	162	141	131	161	155	**390**
6:40	158	137	128	157	151	**400**
6:50	154	134	125	153	148	**410**
7:00	150	131	122	149	144	**420**
7:10	147	128	119	146	141	**430**
7:20	144	125	116	143	137	**440**
7:30	140	122	114	139	134	**450**
7:40	137	119	111	136	131	**460**
7:50	134	117	109	133	129	**470**
8:00	132	115	107	131	126	**480**

Minutes and Seconds ➤

➤ *Seconds*

GROUP THREE

Unit ➤	11	12	13	14	15	
No. of Words ➤	846	811	928	871	1094	
1:30	564	541	619	581	729	90
1:40	508	487	557	523	656	100
1:50	461	442	506	475	597	110
2:00	423	406	464	436	547	120
2:10	390	374	428	402	505	130
2:20	363	348	398	373	469	140
2:30	338	324	371	348	438	150
2:40	317	304	348	327	410	160
2:50	299	286	328	307	386	170
3:00	282	270	309	290	365	180
3:10	267	256	293	275	345	190
3:20	254	243	278	261	328	200
3:30	242	232	265	249	313	210
3:40	231	221	253	238	298	220
3:50	221	212	242	227	285	230
4:00	212	203	232	218	274	240
4:10	203	195	223	209	263	250
4:20	195	187	214	201	252	260
4:30	188	180	206	194	243	270
4:40	181	174	199	187	234	280
4:50	175	168	192	180	226	290
5:00	169	162	186	174	219	300
5:10	164	157	180	169	212	310
5:20	159	152	174	163	205	320
5:30	154	147	169	158	199	330
5:40	149	143	164	154	193	340
5:50	145	139	159	149	188	350
6:00	141	135	155	145	182	360
6:10	137	132	150	141	177	370
6:20	134	128	147	138	173	380
6:30	130	125	143	134	168	390
6:40	127	122	139	131	164	400
6:50	124	119	136	127	160	410
7:00	121	116	133	124	156	420
7:10	118	113	129	122	153	430
7:20	115	111	127	119	149	440
7:30	113	108	124	116	146	450
7:40	110	106	121	114	143	460
7:50	108	104	118	111	140	470
8:00	106	101	116	109	137	480

Minutes and Seconds ➤

◄ Seconds

READING SPEED

Directions: *Write your Words-per-Minute score for each unit in the box under the number of the unit. Then plot your reading speed on the graph by putting a small* **x** *on the line directly above the number of the unit, across from the number of words per minute you read. As you mark your speed for each unit, graph your progress by drawing a line to connect the*

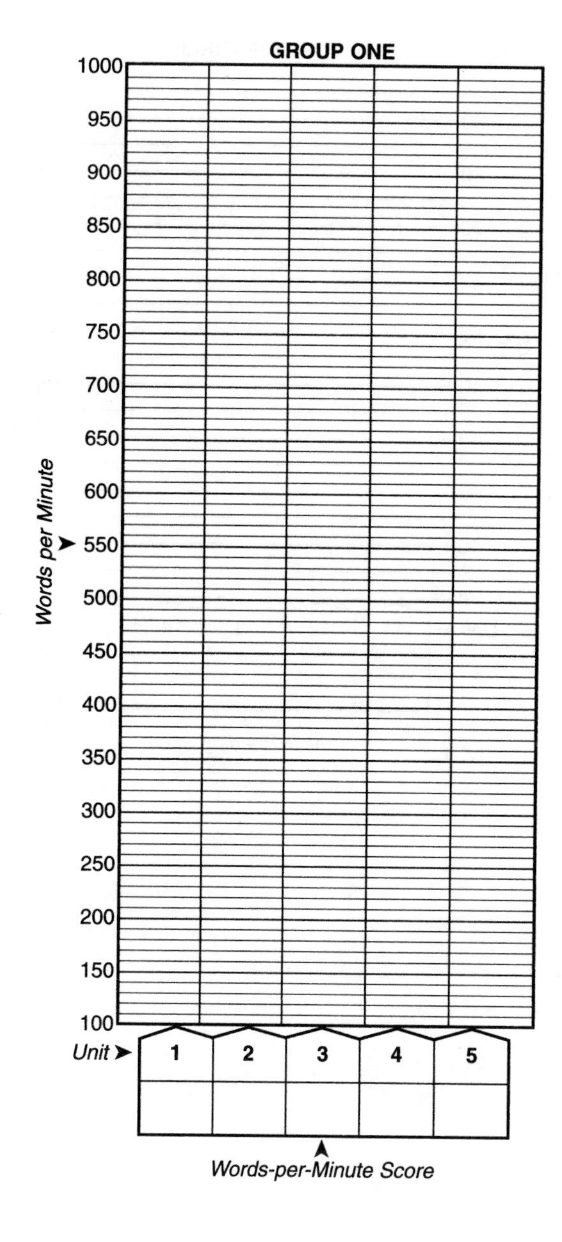

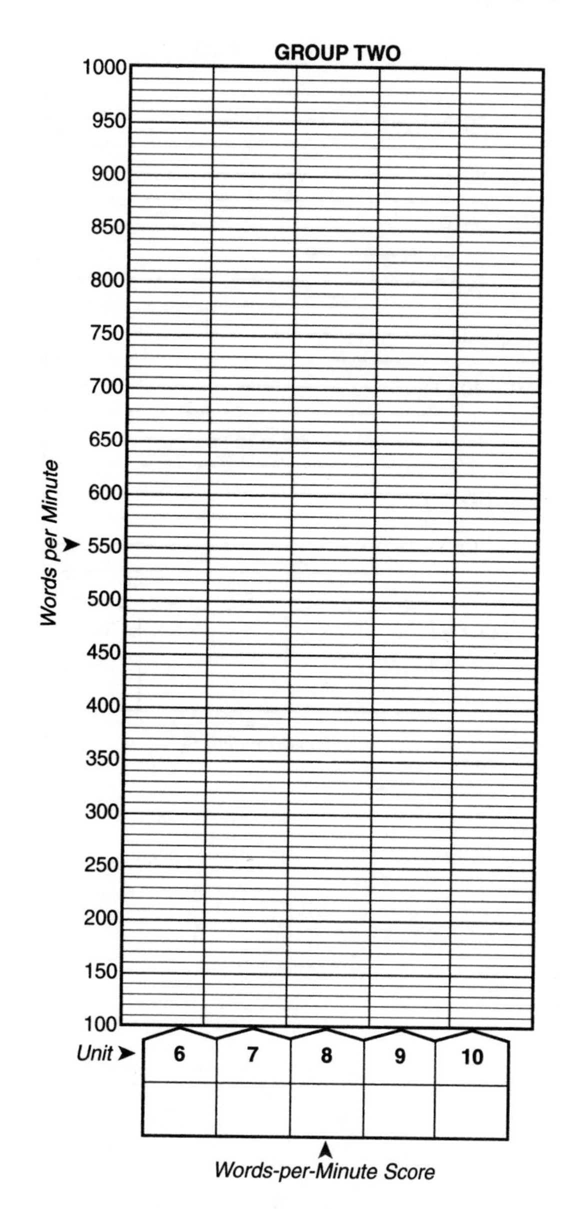

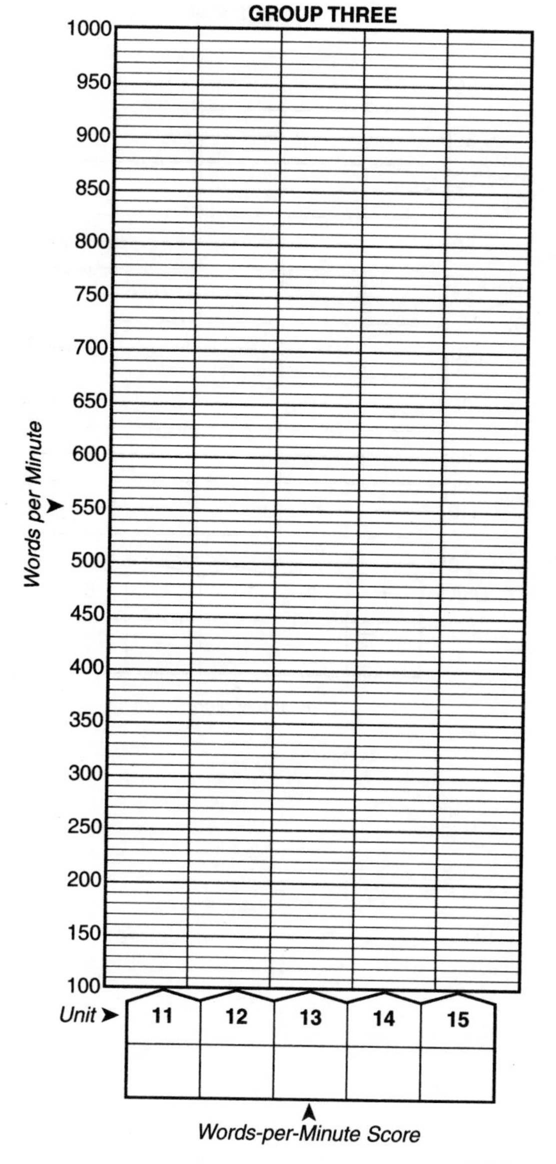

CRITICAL READING SCORES

Directions: *Write your Critical Reading Score for each unit in the box under the number of the unit. Then plot your score on the graph by putting a small* **x** *on the line directly above the number of the unit, across from the score you earned. As you mark your score for each unit, graph your progress by drawing a line to connect the* **x***'s.*

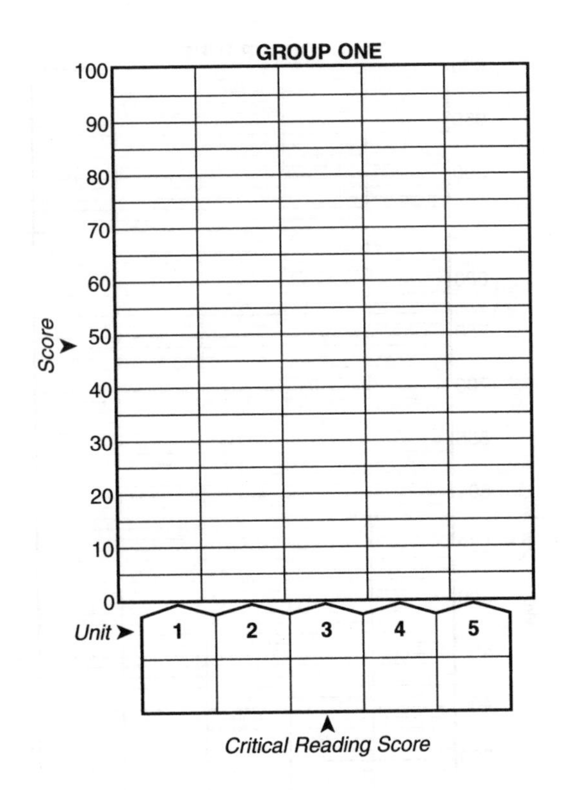

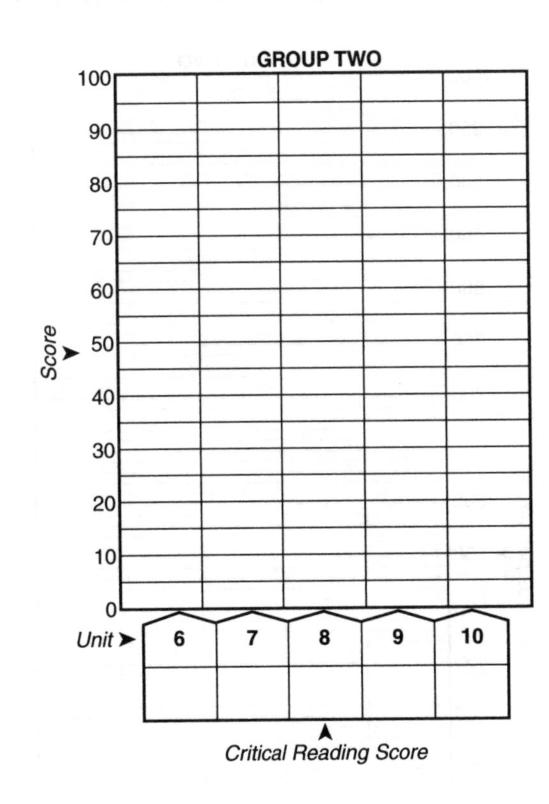

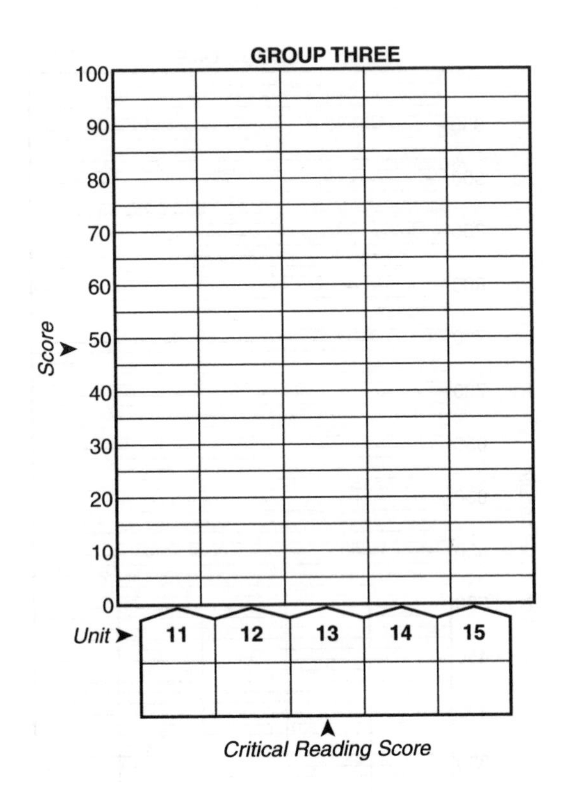